农林英语阅读与翻译

READING AND TRANSLATION FOR AGRICULTURE AND FORESTRY

李 清 编著

合肥工业大学出版社

前　言

作为安徽农业大学外国语学院一名英语教师，作者除了日常教学与科研工作之外，多年来也一直承担本校新农村发展研究院国际合作农业推广口笔译工作。作者主讲翻译类课程，组织并指导学生进行翻译实践，承担本校相关国际会议和国际交流中各种翻译任务。

本书融相关翻译理论（技巧）与实践为一体，能通过大量有针对性的翻译与练习指导学生并藉此提高学生相关翻译能力。本书涵盖涉农林笔译与口译各方面，包括农业政策、农林科普、国际交流和文化经典四大板块，共十六个单元。本书结合当今语言学习和应用实际，有助于培养并加强译者的跨文化交际意识。本书也是对现有相关翻译教学、研究和实践进行补充、拓展乃至创新的尝试。

感谢许有江教授、朱跃教授对作者多年的指导、鼓励和鞭策！

本书为安徽省省级教学团队"翻译教学团队"项目（2019jxtd035）的建设成果。此外，本书出版还得到农业部和财政部重大农技推广服务试点项目（AAU－CSU农业推广与经济发展联合研究院联合研发项目）——"美国大学推广体系重要文献翻译与编印"（财农〔2015〕64号）、安徽农业大学2019年校级质量工程项目"精品线下开放课程——农林英语阅读与翻译"（校教字〔2019〕40号）的资金支持。作者在此一并表示感谢！

感谢校内外所有相关部门和人员的大力支持！

恳请各位批评指正！谢谢！

李　清

2019年9月19日

于安徽农业大学外国语学院

目 录

第一部分 农业政策
PART ONE AGRICULTURAL POLICIES

第一单元 2030 粮食和农业议程（节选）
UNIT ONE FOOD AND AGRICULTURE IN THE 2030 AGENDA
　　　　（EXCERPTED） ………………………………… （002）

第二单元 中国的中医药（节选）
UNIT TWO TRADITIONAL CHINESE MEDICINE (TCM IN CHINA)
　　　　（EXCERPTED） ………………………………… （016）

第三单元 2018 年政府工作报告（节选）
UNIT THREE REPORT ON THE WORK OF THE GOVERNMENT
　　　　IN 2018（EXCERPTED） ……………………… （045）

第四单元 中华人民共和国国民经济和社会发展第十三个五年规划纲要（节选）
UNIT FOUR THE 13TH FIVE-YEAR PLAN FOR ECONOMIC AND
　　　　SOCIAL DEVELOPMENT OF THE PEOPLE'S REPUBLIC
　　　　OF CHINA（EXCERPTED） ……………………… （061）

第二部分 农林科普
PART TWO POPULAR SCIENCES OF AGRICULTURE AND FORESTRY

第一单元 研发超级稻
UNIT ONE THE SUPER RICE CHALLENGE ………………… （079）

第二单元 生命物质的生物化学
UNIT TWO BIOCHEMISTRY OF LIVING MATTER ………… （090）

第三单元　论文摘要
UNIT THREE　ABSTRACTS OF PAPERS ……………… (098)

第四单元　中国农业推广体系的演变与发展（节选）
UNIT FOUR　EVOLUTION AND DEVELOPMENT OF AGRICULTURAL EXTENSION SYSTEM IN CHINA (EXCERPTED)
……………………………………………………………… (124)

第三部分　国际交流
PART THREE　INTERNATIONAL EXCHANGE

第一单元　科罗拉多州前沿区域农业价值链创新集群的出现（节选）
UNIT ONE　THE EMERGENCE OF AN INNOVATION CLUSTER IN THE AGRICULTURAL VALUE CHAIN ALONG COLORADO'S FRONT RANGE (EXCEPTED) ……………… (132)

第二单元　安徽农业大学简介
UNIT TWO　A BRIEF INTRODUCTION TO ANHUI AGRICULTURAL UNIVERSITY ……………………………… (148)

第三单元　共谋绿色生活，共建美丽家园
UNIT THREE　WORKING TOGETHER FOR A GREEN AND BETTER FUTURE FOR ALL ……………… (160)

第四单元　齐心开创共建"一带一路"美好未来
UNIT FOUR　WORKING TOGETHER TO DELIVER A BRIGHTER FUTURE FOR BELT AND ROAD COOPERATION
……………………………………………………………… (175)

第四部分　文化经典
PART FOUR　CULTURAL CLASSICS

第一单元　农舍（节选）
UNIT ONE　ON COTTAGES IN GENERAL (EXCERPTED)
……………………………………………………………… (194)

第二单元　荷塘月色
UNIT TWO　MOONLIGHT OVER THE LOTUS POND ……… (205)

第三单元　清明（诗歌）
UNIT THREE　THE QINGMING FESTIVAL（POEM） ……… (214)

第四单元　秋水（节选）
UNIT FOUR　THE FLOODS OF AUTUMN（EXCERPTED）
　……………………………………………………………… (223)

参考答案 ……………………………………………………… (239)

参考文献 ……………………………………………………… (241)

第一部分　农业政策

PART ONE　AGRICULTURAL POLICIES

第一单元 2030粮食和农业议程（节选）
UNIT ONE
FOOD AND AGRICULTURE IN THE 2030 AGENDA (EXCERPTED)

Focusing on food and agriculture, investing in family farmers and transforming the rural sector can spur progress towards SDG (Sustainable Development Goal) targets. Around the world, over 820 million—one in nine of the world's population—are still chronically undernourished. Among children, more than one in five is stunted.

以粮食和农业为重点，对家庭农户进行投资并改造农村部门，可以促进实现可持续发展目标的进展。在世界各地，超过8.2亿人（占世界人口的九分之一）仍然长期处于营养不良。在儿童中，超过五分之一的儿童发育不良。

But malnutrition is not about hunger alone: poor or unhealthy diets are causing micronutrient deficiencies. Individual and public health is reeling from an unfolding obesity epidemic and associated non-communicable diseases. In nation after nation, the repercussions of climate change are compromising development gains, further fueling tensions in conflict zones. Unstructured urbanization degrades ecosystems while failing to fulfil the promise of prosperity for all. And as resource depletion threatens our planet's continued viability, humanity seems willing to further mortgage its future through irreversible biodiversity loss.

而营养不良不仅仅是和饥饿有关：不良或不健康的饮食会导致微量营养素缺乏。个人和公共卫生正受到不断蔓延的肥胖症和相关非传染性疾病的影响。在一个又一个国家，气候变化的影响正逐步损害发展成果，进一步加剧冲突地区的紧张局势。无条理的城市化使得生态系统退化，人人享有繁荣的承诺无法实现。资源枯竭威胁着我们星球的持续生存，人类似乎愿意通过不可逆转的生物多样性的丧失来进一步抵押其未来的发展。

From this alarming perspective, the rural poor—four-fifths of all of those living in poverty—may appear as another dispiriting dataset. It takes vision and courage to see them as an untapped resource. Socially left behind and afflicted by some of the worst nutritional indicators, they nonetheless supply 80 percent of the world's food. In many countries, agriculture remains the largest employer and main economic sector—a major problem and major opportunity rolled into one.

从这个令人担忧的角度来看，农村贫困人口——占所有贫困人口的五分之四——可能会成为另一个令人沮丧的数据集。将他们视为一种尚未开发的资源需要远见和勇气。尽管他们被社会抛在后面，并受到一些最糟糕的营养指标的影响，但他们仍然提供了世界上80%的食物。在许多国家，农业仍然提供了最多的工作岗位，农业部也仍是主要的经济部门——集重大问题和重大机会于一体。

To achieve the SDGs, we must imperatively make it less of the former and more of the latter. The 2030 Agenda recognizes the fundamental connection between people and planet, agriculture and sustainability. This awareness must urgently be translated into investment in rural people, family farmers, fishing communities, foresters and pastoralists; in food systems that are better balanced, more equitable and less wasteful; in agricultural innovation; and in an approach to natural resources that allies environmental concerns to the pursuit of food security and decent livelihoods for all.

为了实现可持续发展目标，我们必须做到分清轻重缓急。"2030议程"认识到人类与地球、农业与可持续发展之间的根本联系。而这种认识必须立即转化为对农村人口、农民家庭、渔业社区、林业工作者和牧民的投资投入；对粮食系统中更平衡、更公平和更少浪费的投资投入；对农业创新的投资投入；对农业自然资源的投资投入，将对环境的种种关切问题与追求粮食安全和所有人的体面生计结合起来。

ZERO IN ON ZERO HUNGER
实现零饥饿

With the number of those chronically malnourished still dauntingly high, Zero Hunger is at risk. But renewed commitment is essential to realize the vision that inspired the SDGs. Achieving the 2030 Agenda calls for a redoubling of efforts. Now is the time to scale up actions already underway.

由于长期营养不良的人数仍然高得令人畏惧，实现零饥饿目标亦正风险当头。但是，更新的投入对于实现激发可持续发展目标的愿景至关重要。实现"2030议程"需要加倍努力。现在是升级强化当下行动的时候了。

Responding to the spirit of global solidarity to overcome common challenges, countries are broadening participation, forging new partnerships and calling for the involvement of all development actors—the United Nations system, civil society, the private sector, the donor community, academia, cooperatives, and others.

为了响应全球团结精神以克服共同挑战，各国正在扩大参与范围，建立新型伙伴关系，并呼吁所有致力于发展的行动者——联合国系统、民间社会、私营部门、捐助界、学术界、合作社和其他机构的参与。

National platforms are being established to develop more integrated programmes and policies, to better interlink different goals and targets. At the same time, multi-stakeholder mechanisms and new forms of participatory governance structures are bolstering policy ownership, while helping to mobilize capacities, information, technologies, and access to financial and production resources.

国家平台正在建立以制定更综合统筹的方案和政策，从而更好地将不同的目的和目标联合起来。与此同时，多个利益攸关方机制和新型参与性治理结构正在强化政策所有权，同时帮助调动各方能力、信息、技术并获得财政和生产资源。

As international cooperation increases, UN support to countries is shifting to emphasize policy advice, institutional capacity development and monitoring of progress. The role of the UN system as a trusted and neutral facilitator of support processes and partnerships is fundamental.

随着国际合作的加强，联合国对各国的支持正在转向强调政策建议、机构能力建设和进展监测等方面。联合国系统作为支持各项进程和伙伴关系的可信的、中立的推动者作用至关重要。

FAO has identified multi-stakeholder partnerships as one of the key drivers of its support to the 2030 Agenda. The Organization plays a leading role in governance matters and participatory approaches to policy-making, bringing together diverse state and non-state players to interact and discuss policy, supplying essential data, norms and standards, and supporting countries in implementing inclusive and cross-cutting actions. At global, regional and national levels, FAO builds partnerships to support enabling environments for policies and programmes to achieve transformative change on food security and nutrition and sustainable agriculture. The Organization works to strengthen the capacities of stakeholders and mobilize resources in order to accelerate efforts aimed at rural transformation and ending poverty and hunger.

联合国粮农组织已将多个利益攸关方伙伴关系确定为其支持"2030议程"的主要动力之一。本组织在治理事务和政策制定参与方面发挥主导作用，能汇集不同的国家和非国家参与者，相互交流并讨论政策，提供必要的数据、规范和标准，支持各国实施包容和跨领域行动。在全球、区域和国家各级，粮农组织建立各种伙伴关系，为政策和方案创造有利环境，以实现粮食安全、营养和可持续农业方面的变革。本组织致力于加强利益攸关方的各种能力，调动各方资源，从而加速针对农村转型和消除贫困与饥饿的种种努力。

HEALTHY PEOPLE, HEALTHY PLANET
健康的人民，健康的地球

Development at the expense of the environment is no longer an option for a growing global population. The MDG (Millennium Development Goals) era, 1990—2015, brought significant benefits to millions of people, including by nearly halving the proportion of hungry people in the world. However, much of humanity's progress has come at a considerable cost to the environment. High-input, resource-intensive farming has contributed to deforestation, water scarcity, soil depletion and high levels of greenhouse gas emissions. Today, there are more people on our planet to feed with less water and productive land.

以牺牲环境为代价的发展已不再是日益增长的全球人口的一种选择。1990年至2015年的千年发展目标时代为数百万人带来了巨大的好处,其中包括使世界饥饿人口的比例减少了近一半。然而,人类的进步在很大程度上是以牺牲环境为巨额代价的。高投入、资源密集的农业导致了森林砍伐、水资源短缺、土壤枯竭和温室气体的高额排放。今天,地球上有更多的人,而水更少,多产的土地也更少。

The 2030 Agenda formalizes the need to conserve natural resources and biodiversity: they are to be managed responsibly, in the interest of humankind and the planet that sustains it. Agriculture and food systems must be transformed away from wasteful, energy-hungry and exploitative approaches. Policymakers and partners should encourage sustainable production and consumption patterns, while ensuring higher governance standards and more inclusive decision-making.

"2030议程"明确了保护自然资源和生物多样性的必要性:为了人类和维持自然资源和生物多样性的地球的利益,必须负责任地管理自然资源和生物多样性。必须改变农业和粮食系统,使其摆脱浪费、高耗能和资源开发的做法。政策制定者和合作伙伴应鼓励可持续的生产和消费模式,同时确保更高的治理标准和更具包容性的决策。

TIME FOR A GREENER REVOLUTION
践行绿色革命

The Green Revolution of the mid-to-late 20th century provided a much needed increase in agricultural productivity to keep pace with rapid population growth. It is now time for a second green revolution in which agriculture continues to provide abundant and healthy food while at the same time promoting the conservation and use of ecosystem services and biodiversity. The potential exists to reverse the trends that lead to natural resources degradation, deforestation, salinization of soils and desertification. Approaches exist to produce more and healthier food in a sustainable way with fewer resources, reducing encroachment on natural ecosystems, including forests and wetlands.

20世纪中后期的绿色革命为农业生产率提供了一个急需的提升，以跟上人口的快速增长。现在是进行第二次绿色革命的时候了，在这场革命中，农业继续提供丰富和健康的食物，同时促进生态系统服务和生物多样性的保护和利用。这有可能扭转导致自然资源退化、森林砍伐、土壤盐碱化和沙漠化的趋势。现有的方法是用更少的资源以可持续发展的方式生产更多和更健康的食品，减少对包括森林和湿地在内的自然生态系统的侵蚀。

A COMMON VISION FOR SUSTAINABLE FOOD AND AGRICULTURE
可持续发展的粮食和农业之共同愿景

Balancing the different dimensions of sustainability is at the heart of FAO's Common Vision for Food and Agriculture. Working with partners, FAO has developed sustainable approaches in areas such as agroecology, agroforestry, biotechnology, and climate-smart and conservation agriculture that bring together traditional knowledge, modern technology and innovation. Capacity development supports their adaptation at community and country levels to ensure local relevance and applicability.

平衡可持续性发展的方方面面是粮农组织粮食和农业共同愿景的核心。粮农组织与伙伴合作，在农业生态学、农林业、生物技术、气候智能和保护农业等领域制订了可持续发展方案，将传统知识、现代技术和创新结合起来。能力发展则支持它们在社区和国家级别的适应，以确保当地相关性和适用性。

WEATHERING A CHANGING CLIMATE
经受住气候变化

Over the past decade or so, the agriculture sector has absorbed at least

25 percent of the total damage and losses caused by droughts, floods and storms and other climate extreme events. Those who are now suffering most have contributed least to the changing climate. Farmers, pastoralists, fisherfolk and community foresters depend on activities that are intimately and inextricably linked to climate. They will require greater access to technologies, markets, information and credit for investment to adapt their production systems and practices.

在过去十年左右的时间里,农业部门至少承受了干旱、洪水和风暴等极端气候事件造成的总损失的25%。那些现在受苦最多的人们对气候变化的影响最小。农民、牧民、渔民和社区林业工作者都依赖与气候密切相关的各种活动。他们将需要获得更多的技术、市场、信息和信贷,以便投资来调整其生产系统和实践。

The climate is changing. Agriculture must too. Food production threatens to be the greatest casualty of climate change, but sustainable agriculture has the ability, through adaptation and mitigation, to respond to more extreme weather events.

气候正在变化,农业也必须改变。粮食生产可能成为气候变化的最大受害者,但可持续农业有能力通过适应和缓解方式来应对更多的极端天气事件。

Climate change is having profound consequences on our planet's diversity of life and on people's lives. Oceans are warming. Sea levels are rising, creating an existential threat for dozens of small island states. Longer, more intense droughts threaten freshwater supplies and crops, endangering efforts to feed a growing world population.

Without action, the changing climate will seriously compromise food production in countries and regions that are already highly food insecure. It will affect food availability by reducing the productivity of crops, livestock and fisheries, and hinder access to food by disrupting the livelihoods of millions of rural people who depend on agriculture for their incomes. It will expose both urban and rural poor to higher and more volatile food prices.

Ultimately, it will jeopardize progress towards the SDGs.

气候变化正在对我们星球上的生命多样性和人们的生活产生深远的影响。海洋正在变暖。海平面正在上升，对数十个小岛屿国家构成生存威胁。持续时间更长、程度更严重的干旱威胁着淡水供应和农作物产出，危及日益增长的世界人口的生存。如果不采取行动，不断变化的气候将严重损害已经高度缺乏粮食保障的国家和地区的粮食生产。它将通过降低农作物、牲畜和渔业的生产率来影响粮食供应，并通过破坏依赖农业获得收入的无数农民的生计来阻碍粮食供应。这将使城市和农村的穷人都面临更高和更不稳定的食品价格。最终，这将危及可持续发展目标的进展。

Agriculture has a major role to play in responding to climate change. FAO is now supporting countries to both adapt to and mitigate the effects of climate change through research-based programs and projects, with a focus on adapting smallholder production and making the livelihoods of rural populations more resilient.

农业在应对气候变化方面可发挥重要作用。粮农组织目前正在通过以研究为基础的方案和项目支持各国适应和减轻气候变化的影响，重点是调整小农生产，使农村人口的生计更具弹性。

Tied to the principle of leaving no one behind, and driven by means of partnerships and accountability, FAO's broad priorities in the 2030 Agenda are to:

1. End poverty, hunger and all forms of malnutrition
2. Enable sustainable development in agriculture, fisheries and forestry
3. Respond to climate change and build resilient communities.

结合不落下一个人的原则，由各伙伴关系和各责任措施驱动，粮农组织"2030议程"覆盖面广泛的优先级目标是：

1. 结束贫穷、饥饿和各种形式的营养不良；
2. 赋予农业、渔业和林业之可持续发展；
3. 应对气候变化并建立弹性恢复机制。

注解
Notes

THE 17 SUSTAINABLE DEVELOPMENT GOALS
17 个可持续发展目标

GOAL 1　End poverty in all its forms everywhere
目标 1　消除世界各地各种形式的贫困

GOAL 2　End hunger, achieve food security and improved nutrition and promote sustainable agriculture
目标 2　消除饥饿，实现粮食安全和改善营养，促进可持续农业

GOAL 3　Ensure healthy lives and promote well-being for all at all ages
目标 3　确保所有人的健康生活，促进所有年龄层的福祉

GOAL 4　Ensure inclusive and quality education for all and promote lifelong learning
目标 4　确保全民接受优质教育，促进终身学习

GOAL 5　Achieve gender equality and empower all women and girls
目标 5　实现性别平等，赋予所有妇女和女童权力

GOAL 6　Ensure access to water and sanitation for all
目标 6　确保人人享有水和卫生设施

GOAL 7　Ensure access to affordable, reliable, sustainable and modern energy for all
目标 7　确保所有人都能获得负担得起、可靠、可持续和现代化的能源

GOAL 8　Promote inclusive and sustainable economic growth, employment and decent work for all

目标 8 　　促进包容的可持续经济增长、就业且人人享有体面工作

GOAL 9 　Build resilient infrastructure, promote sustainable industrialization and foster innovation

目标 9 　　建设有韧性的基础设施，促进可持续工业化并促进创新

GOAL 10 　Reduce inequality within and among countries

目标 10 　　减少国家内部和国家之间的不平等

GOAL 11 　Make cities inclusive, safe, resilient and sustainable

目标 11 　　建设包容、安全、有韧性、可持续的城市

GOAL 12 　Ensure sustainable consumption and production patterns

目标 12 　　确保可持续的消费和生产模式

GOAL 13 　Take urgent action to combat climate change and its impacts

目标 13 　　采取紧急行动应对气候变化及其影响

GOAL 14 　Conserve and sustainably use the oceans, seas and marine resources

目标 14 　　养护和可持续利用种种海洋资源

GOAL 15 　Sustainably manage forests, combat desertification, halt and reverse land degradation, halt biodiversity loss

目标 15 　　可持续管理森林，防治荒漠化，制止和扭转土地退化，制止生物多样性丧失

GOAL 16 　Promote just, peaceful and inclusive societies

目标 16 　　促进建设公正、和平、包容的社会

GOAL 17 　Revitalize the global partnership for sustainable development

目标 17　重振全球可持续发展伙伴关系

(Excerpted from "FOOD AND AGRICULTURE IN THE 2030 AGENDA" www. fao. org/3/ca5299en/ca5299en. pdf 2019 - 07 - 17)

练习
Exercises

练习一　将原课文内容做笔译和视译

练习二　词汇和语法

VOCABULARY AND STRUCTURE

There are 35 sentences in this section. Beneath each sentence there are four words or phrases marked A, B, C, and D. Choose one word or phrase that best completes the sentence.

1. In the past few years the company has _____ a lot of money training its staff in computer science.
　　A. spent　　　　B. cost　　　　C. paid　　　　D. taken

2. The noise was so _____ that only those with excellent hearing were aware of it.
　　A. dim　　　　B. soft　　　　C. faint　　　　D. gentle

3. The worth of intelligence _____ the society people live in.
　　A. determines　　　　　　　　B. is determined by
　　C. is decided on　　　　　　　D. decides

4. Some people either _____ avoid questions of right and wrong or remain neutral about them.
　　A. violently　　　　　　　　B. sincerely
　　C. properly　　　　　　　　D. deliberately

5. _____ a good beginning is made, the work is half done.
　　A. As soon as　　B. While　　C. As　　　　D. Once

6. We shall probably never be able to _____ the exact nature of these sub-atomic particles.

A. assert B. impart C. ascertain D. notify

7. It was _____ that the restaurant discriminated against black customers.

A. addicted B. alleged C. assaulted D. ascribed

8. Many years had _____ before they returned to their original urban areas.

A. floated B. elapsed
C. skipped D. proceeded

9. What you say now is not _____ with what you said last week.

A. consistent B. persistent C. permanent D. insistent

10. Military orders are _____ and cannot be disobeyed.

A. defective B. conservative
C. alternative D. imperative

11. It is clear that the whole world is passing through a social revolution in which a central _____ must be taken by scientists and technologists.

A. process B. attention C. measure D. part

12. The farmers were more anxious for rain than the people in the city because they had more _____.

A. at length B. at last C. at stake D. at ease

13. When the big bills for Mother's hospital care came, Father was glad he had money in the bank to _____.

A. fall short of B. fall through C. fall back on D. fall in with

14. These plastic flowers look so _____ that many people think they are real.

A. beautiful B. natural C. artificial D. similar

15. The managing director promised that he would _____ me as soon as he had any further information.

A. communicate B. notice C. notify D. note

16. They believed that this was not the _____ of their campaign for equality but merely the beginning.

A. climax B. summit
C. pitch D. maximum

17. Several guests waiting in the _____ for the front door to open.
   A. porch B. vent C. inlet D. entry

18. As the mountains were covered with a _____ of cloud, we couldn't see their tops.
   A. coating B. film C. veil D. shade

19. We couldn't really afford to buy a house so we got it on hire purchase and paid monthly _____.
   A. investments B. requirements
   C. arrangements D. installments

20. The magician made us think he cut the girl into pieces but it was merely an _____.
   A. illusion B. impression C. image D. illu

21. _____ it is you've found, you must give it back to the owner.
   A. That B. What C. Whatever D. Though

22. _____ arriving at the station, the Queen was welcomed by the people there.
   A. At B. For C. On D. In

23. _____ the electric current pass through conductors that it is difficult for us to imagine its speed.
   A. So does fast B. Such does fast
   C. So fast does D. Such fast does

24. Prince Henry lived in the fifteenth century. As a boy he became devoted to the sea, and he dedicated himself to _____ the design of ships and the methods of sailing them.
   A. improving B. being improved
   C. improve D. improved

25. _____ oxygen to every part of the body, the blood comes back to the right side of the heart.
   A. Delivering B. Having delivered
   C. Having been delivered D. To deliver

26. _____ I love you, I cannot let you have any more money.

A. Much as   B. Whether

C. Also   D. However

27. There are many forms of energy, _____ is atomic energy.

A. one of that   B. one of whom

C. which of one   D. one of which

28. If I hadn't _____ by an expert, it wouldn't look so nice.

A. had it done   B. done it

C. have it done   D. have done it

29. I must go now. _____, if you want that book, I'll bring it next time.

A. Occasionally   B. Accidentally

C. Incidentally   D. Subsequently

30. _____ a tie for first place in the competition, a runoff will be held.

A. Should there be   B. If there had been

C. Was there   D. Could there be

31. There is no one of us _____ wishes to go.

A. but   B. who   C. that   D. unless

32. His ideas were far _____ the age in which he lived.

A. in front of   B. in adventure of

C. in advantage of   D. in advance of

33. The Minister is pleased to be able to announce that another 500 miles of motorway _____ by the end of next year.

A. will be building   B. are building

C. have been built   D. will have been built

34. _____ do the documents indicate any degree of difference in the legal status of husband or wife.

A. In any case   B. In that case   C. In no case   D. In case

35. Rachel is _____ courageous than Saul is.

A. no more   B. not so

C. many more   D. not much

第二单元　中国的中医药（节选）
UNIT TWO
TRADITIONAL CHINESE MEDICINE (TCM IN CHINA)
(EXCERPTED)

前 言
Preface

人类在漫长发展进程中创造了丰富多彩的世界文明，中华文明是世界文明多样性、多元化的重要组成部分。中医药作为中华文明的杰出代表，是中国各族人民在几千年生产生活实践和与疾病作斗争中逐步形成并不断丰富发展的医学科学，不仅为中华民族繁衍昌盛作出了卓越贡献，也对世界文明进步产生了积极影响。

Humanity has created a colorful global civilization in the long course of its development, and the civilization of China is an important component of the world civilization harboring great diversity. As a representative feature of Chinese civilization, traditional Chinese medicine (TCM) is a medical science that was formed and developed in the daily life of the people and in the process of their fight against diseases over thousands of years. It has made a great contribution to the nation's procreation and the country's prosperity, in addition to producing a positive impact on the progress of human civilization.

中医药在历史发展进程中，兼容并蓄、创新开放，形成了独特的生命观、健康观、疾病观、防治观，实现了自然科学与人文科学的融合和统一，蕴含了中华民族深邃的哲学思想。随着人们健康观念变化和医学模式转变，中医药越来越显示出独特价值。

TCM has created unique views on life, on fitness, on diseases and on the prevention and treatment of diseases during its long history of absorption and innovation. It represents a combination of natural sciences and humanities, embracing profound philosophical ideas of the Chinese nation. As ideas on fitness and medical models change and evolve, traditional Chinese medicine has come to underline a more and more profound value.

新中国成立以来,中国高度重视和大力支持中医药发展。中医药与西医药优势互补,相互促进,共同维护和增进民众健康,已经成为中国特色医药卫生与健康事业的重要特征和显著优势。

Since the founding of the People's Republic of China in 1949, the Chinese government has set great store by TCM and rendered vigorous support to its development. TCM and Western medicine have their different strengths. They work together in China to protect people from diseases and improve public health. This has turned out to be one of the important features and notable strengths of medicine with Chinese characteristics.

一、中医药的历史发展
Ⅰ. The Historical Development of TCM

1. 中医药历史发展脉络
1. History of TCM

在远古时代,中华民族的祖先发现了一些动植物可以解除病痛,积累了一些用药知识。随着人类的进化,人们开始有目的地寻找防治疾病的药物和方法,所谓"神农尝百草""药食同源",就是当时的真实写照。夏代(约前2070—前1600)酒和商代(前1600—前1046)汤液的发明,为提高用药效果提供了帮助。

In remote antiquity, the ancestors of the Chinese nation chanced to find that some creatures and plants could serve as remedies for certain ailments and pains, and came to gradually master their application. As time went by, people began to actively seek out such remedies and methods for preventing and treating diseases. Sayings like "Shennong (Celestial Farmer) tasting a hundred herbs" and "food and medicine coming from the same source" are characteristic of those years. The discovery of alcohol in the Xia Dynasty (c. 2070—1600 BC) and the invention of herbal decoction in the Shang Dynasty (1600—1046 BC) rendered medicines more effective.

进入西周时期（前 1046—前 771），开始有了食医、疾医、疡医、兽医的分工。春秋战国（前 770—前 221）时期，扁鹊总结前人经验，提出"望、闻、问、切"四诊合参的方法，奠定了中医临床诊断和治疗的基础。

In the Western Zhou Dynasty (1046—771 BC), doctors began to be classified into four categories-dietician, physician, doctor of decoctions and veterinarian. During the Spring and Autumn and Warring States Period (770—221 BC), Bian Que drew on the experience of his predecessors and put forward the four diagnostic methods-inspection, auscultation & olfaction, inquiry, and palpation, laying the foundation for TCM diagnosis and treatment.

秦汉时期（前 221—公元 220）的中医典籍《黄帝内经》，系统论述了人的生理、病理、疾病以及"治未病"和疾病治疗的原则及方法，确立了中医学的思维模式，标志着从单纯的临床经验积累发展到了系统理论总结阶段，形成了中医药理论体系框架。

The *Huang Di Nei Jing* (Yellow Emperor's Inner Canon) compiled during the Qin and Han times (221 BC—AD 220) offered systematic discourses on human physiology, on pathology, on the symptoms of illness, on preventative treatment, and on the principles and methods of treatment. This book defined the framework of TCM, thus serving as a landmark in TCM's development and symbolizing the transformation from the accumulation of clinical experience to the systematic summation of theories. A theoretical framework for TCM had been in place.

东汉时期，张仲景的《伤寒杂病论》，提出了外感热病（包括瘟疫等传染病）的诊治原则和方法，论述了内伤杂病的病因、病证、诊法、治疗、预防等辨证规律和原则，确立了辨证论治的理论和方法体系。

The *Shang Han Za Bing Lun* (Treatise on Febrile Diseases and

Miscellaneous Illnesses) collated by Zhang Zhongjing in the Eastern Han Dynasty (25—220) advanced the principles and methods to treat febrile diseases due to exogenous factors (including pestilences). It expounds on the rules and principles of differentiating the patterns of miscellaneous illnesses caused by internal ailments, including their prevention, pathology, symptoms, therapies, and treatment. It establishes the theory and methodology for syndrome pattern diagnosis and treatment differentiation.

同时期的《神农本草经》，概括论述了君臣佐使、七情合和、四气五味等药物配伍和药性理论，对于合理处方、安全用药、提高疗效具有十分重要的指导作用，为中药学理论体系的形成与发展奠定了基础。

The *Shen Nong Ben Cao Jing* (Shennong's Classic of Materia Medica) —another masterpiece of medical literature appeared during this period—outlines the theory of the compatibility of medicinal ingredients. For example, it holds that a prescription should include at the same time the jun (or sovereign), chen (or minister), zuo (or assistant) and shi (or messenger) ingredient drugs, and should give expression to the harmony of the seven emotions as well as the properties of drugs known as "four natures" and "five flavors." All this provides guidance to the production of TCM prescriptions, safe application of TCM drugs and enhancement of the therapeutic effects, thus laying the foundation for the formation and development of TCM pharmaceutical theory.

东汉末年，华佗创制了麻醉剂"麻沸散"，开创了麻醉药用于外科手术的先河。西晋时期（265—317），皇甫谧的《针灸甲乙经》，系统论述了有关脏腑、经络等理论，初步形成了经络、针灸理论。

In the late years of the Eastern Han Dynasty, Hua Tuo (c. 140—208) was recorded to be the first person to use anesthetic (mafeisan) during surgery. The *Zhen Jiu Jia Yi Jing* (AB Canon of Acupuncture and Moxibustion) by Huangfu Mi during the Western Jin time (265—317) expounded on the concepts of zangfu (internal organs) and jingluo (meridians & collaterals). This was the point when theory of jingluo and acupuncture & moxibustion began to take shape.

唐代（618—907），孙思邈提出的"大医精诚"，体现了中医对医道精微、心怀至诚、言行诚谨的追求，是中华民族高尚的道德情操和卓越的文明智慧在中医药中的集中体现，是中医药文化的核心价值理念。

Sun Simiao, a great doctor of the Tang Dynasty (618—907), proposed that mastership of medicine lies in proficient medical skills and lofty medical ethics, which eventually became the embodiment of a moral value of the Chinese nation, a core value that has been conscientiously upheld by the TCM circles.

明代（1368—1644），李时珍的《本草纲目》，在世界上首次对药用植物进行了科学分类，创新发展了中药学的理论和实践，是一部药物学和博物学巨著。清代（1616—1911），叶天士的《温热论》，提出了温病和时疫的防治原则及方法，形成了中医药防治瘟疫（传染病）的理论和实践体系。

A herbology and nature masterpiece, the *Ben Cao Gang Mu* (Compendium of Materia Medica) compiled by Li Shizhen in the Ming Dynasty (1368—1644) was the first book in the world that scientifically categorized medicinal herbs. It was a pioneering work that advanced TCM pharmaceutical theory. The *Wen Re Lun* (A Treatise on Epidemic Febrile Diseases) by Ye Tianshi during the Qing Dynasty (1616—1911) developed the principles and methods for prevention and treatment of pestilential febrile diseases. It represents the theory and results of the practice of TCM in preventing and treating such diseases.

清代中期以来，特别是民国时期，随着西方医学的传入，一些学者开始探索中西医药学汇通、融合。

Following the spread of Western medicine in China from the mid-Qing Dynasty, especially during the period of the Republic of China (1912—1949), some TCM experts began to explore ways to absorb the essence of Western medicine for a combination of TCM with Western medicine.

2. 中医药特点
2. Characteristics of TCM

在数千年的发展过程中，中医药不断吸收和融合各个时期先进的科学技术和人文思想，不断创新发展，理论体系日趋完善，技术方法更加丰富，形成了鲜明的特点。

During its course of development spanning a couple of millennia, TCM has kept drawing and assimilating advanced elements of natural science and humanities. Through many innovations, its theoretical base covered more ground and its remedies against various diseases expanded, displaying unique characteristics.

第一，重视整体。中医认为人与自然、人与社会是一个相互联系、不可分割的统一体，人体内部也是一个有机的整体。重视自然环境和社会环境对健康与疾病的影响，认为精神与形体密不可分，强调生理和心理的协同关系，重视生理与心理在健康与疾病中的相互影响。

First, setting great store by the holistic view. TCM deems that the relationship between humans and nature is an interactive and inseparable whole, as are the relationships between humans and the society, and between the internal organs of the human body, so it values the impacts of natural and social environment on health and illness. Moreover, it believes that the mind and body are closely connected, emphasizing the coordination of physical and mental factors and their interactions in the conditions of health and illness.

第二，注重"平"与"和"。中医强调和谐对健康具有重要作用，认为人的健康在于各脏腑功能和谐协调，情志表达适度中和，并能顺应不同环境的变化，其根本在于阴阳的动态平衡。疾病的发生，其根本是在内、外因素作用下，人的整体功能失去动态平衡。维护健康就是维护人的整体功能动态平衡，治疗疾病就是使失去动态平衡的整体功能恢复到协调与和谐状态。

Second, setting great store by the principle of harmony. TCM lays particular stress on the importance of harmony on health, holding that a person's physical health depends on harmony in the functions of the various body organs, the moderate status of the emotional expression, and adaption and compliance to different environments, of which the most vital is the dynamic balance between yin and yang. The fundamental reason for illness is that various internal and external factors disturb the dynamic balance. Therefore, maintaining health actually means conserving the dynamic balance of body functions, and curing diseases means restoring chaotic body functions to a state of coordination and harmony.

第三，强调个体化。中医诊疗强调因人、因时、因地制宜，体现为"辨证论治"。"辨证"，就是将四诊（望、闻、问、切）所采集的症状、体征等个体信息，通过分析、综合，判断为某种证候。"论治"，就是根据辨证结果确定相应治疗方法。中医诊疗着眼于"病的人"而不仅是"人的病"，着眼于调整致病因子作用于人体后整体功能失调的状态。

Third, emphasis on individuality. TCM treats a disease based on full consideration of the individual constitution, climatic and seasonal conditions, and environment. This is embodied in the term "giving treatment on the basis of syndrome differentiation." Syndrome differentiation means diagnosing an illness as a certain syndrome on the basis of analyzing the specific symptoms and physical signs collected by way of inspection, auscultation & olfaction, inquiry, and palpation, while giving treatment means defining the treatment approach in line with the syndrome differentiated. TCM therapies focus on the person who is sick rather than the illness that the patient contracts, i.e., aiming to restore the harmonious state of body functions that is disrupted by pathogenic factors.

第四，突出"治未病"。中医"治未病"核心体现在"预防为主"，重在"未病先防、既病防变、瘥后防复"。中医强调生活方式和健康有着密切关系，主张以养生为要务，认为可通过情志调摄、劳逸适度、膳食合理、起居有常等，也可根据不同体质或状态给予适当干预，以养神健体、培育正气，提高抗邪能力，从而达到保健和防病作用。

Fourth, emphasis on preventative treatment. Preventative treatment is a core belief of TCM, which lays great emphasis on prevention before a disease arises, guarding against pathological changes when falling sick, and protecting recovering patients from relapse. TCM believes that lifestyle is closely related to health, so it advocates health should be preserved in daily life. TCM thinks that a person's health can be improved through emotional adjustment, balanced labor and rest, a sensible diet, and a regular life, or through appropriate intervention in the lifestyle based on people's specific physical conditions. By these means, people can cultivate vital energy to protect themselves from harm and keep healthy.

第五，使用简便。中医诊断主要由医生自主通过望、闻、问、切等方法收集患者资料，不依赖于各种复杂的仪器设备。中医干预既有药物，也有针灸、推拿、拔罐、刮痧等非药物疗法。许多非药物疗法不需要复杂器具，其所需器具（如小夹板、刮痧板、火罐等）往往可以就地取材，易于推广使用。

Fifth, simplicity. TCM doctors diagnose patients through inspection, auscultation & olfaction, inquiry, and palpation. In addition to medication, TCM has many non-pharmacological alternative approaches such as acupuncture and moxibustion, tuina (massage), cupping and guasha (spooning). There is no need for complex equipment. TCM tools, for example, the small splints used in Chinese osteopathy, the spoons used in guasha, or the cups used in cupping therapy, can draw from materials close at hand, so that such treatments can spread easily.

3. 中医药的历史贡献
3. TCM's Contributions

中医药是中华优秀传统文化的重要组成部分和典型代表，强调"道法自然、天人合一""阴阳平衡、调和致中""以人为本、悬壶济世"，体现了中华文化的内核。中医药还提倡"三因制宜、辨证论治""固本培元、壮筋续骨""大医精诚、仁心仁术"，更丰富了中华文化内涵，为中华民族认识和改造世界提供了有益启迪。

TCM is an important component and a characteristic feature of traditional Chinese culture. Applying such principles as "man should observe the law of the nature and seek for the unity of the heaven and humanity," "yin and yang should be balanced to obtain the golden mean," and "practice of medicine should aim to help people," TCM embodies the core value of Chinese civilization. TCM also advocates "full consideration of the environment, individual constitution, and climatic and seasonal conditions when practicing syndrome differentiation and determining therapies," "reinforcing the fundamental and cultivating the vital energy, and strengthening tendons and bones," and "mastership of medicine lying in proficient medical skills and lofty medical ethics," all concepts that enrich Chinese culture and provide an enlightened base from which to study and transform the world.

中医药作为中华民族原创的医学科学,从宏观、系统、整体角度揭示人的健康和疾病的发生发展规律,体现了中华民族的认知方式,深深地融入民众的生产生活实践中,形成了独具特色的健康文化和实践,成为人们治病祛疾、强身健体、延年益寿的重要手段,维护着民众健康。从历史上看,中华民族屡经天灾、战乱和瘟疫,却能一次次转危为安,人口不断增加、文明得以传承,中医药作出了重大贡献。

TCM originated in the Chinese culture. It explains health and diseases from a macro, systemic and holistic perspective. It shows how China perceives nature. As a unique form of medicine, TCM exercises a profound influence on the life of the Chinese people. It is a major means to help the Chinese people maintain health, cure diseases, and live a long life. The Chinese nation has survived countless natural disasters, wars and pestilences, and continues to prosper. In this process, TCM has made a great contribution.

中医药发祥于中华大地,在不断汲取世界文明成果、丰富发展自己的同时,也逐步传播到世界各地。早在秦汉时期,中医药就传播到周边国家,并对这些国家的传统医药产生重大影响。预防天花的种痘技术,在明清时代就传遍世界。

Born in China, TCM has also absorbed the essence of other civilizations, evolved, and gradually spread throughout the world. As early as the Qin and Han dynasties (221 BC—AD 220), TCM was popular in many neighboring countries and exerted a major impact on their traditional medicines. The TCM smallpox vaccination technique had already spread outside of China during the Ming and Qing dynasties (1368—1911).

《本草纲目》被翻译成多种文字广为流传，达尔文称之为"中国古代的百科全书"。针灸的神奇疗效引发全球持续的"针灸热"。抗疟药物"青蒿素"的发明，拯救了全球特别是发展中国家数百万人的生命。同时，乳香、没药等南药的广泛引进，丰富了中医药的治疗手段。

The *Ben Cao Gang Mu* (Compendium of Materia Medica) was translated into various languages and widely read, and Charles Darwin, the British biologist, hailed the book as an "ancient Chinese encyclopedia." The remarkable effects of acupuncture and moxibustion have won it popularity throughout the world. The discovery of qinghaosu (artemisinin, an anti-malaria drug) has saved millions of lives, especially in developing countries. Meanwhile, massive imports of medicinal substances such as frankincense and myrrh have enriched TCM therapies.

二、中国发展中医药的政策措施
II. Policies and Measures on TCM Development

中国高度重视中医药事业发展。新中国成立初期，国家把"团结中西医"作为三大卫生工作方针之一，确立了中医药应有的地位和作用。1978年，中共中央转发卫生部《关于认真贯彻党的中医政策，解决中医队伍后继乏人问题的报告》，并在人、财、物等方面给予大力支持，有力地推动了中医药事业发展。中华人民共和国宪法指出，发展现代医药和我国传统医药，保护人民健康。1986年，国务院成立相对独立的中医药管理部门。

China lays great store by the development of TCM. When the People's Republic was founded in 1949, the government placed emphasis on uniting Chinese and Western medicine as one of its three guidelines for health work, and enshrined the important role of TCM. In 1978, the Communist

Party of China (CPC) Central Committee transmitted throughout the country the Ministry of Health's "Report on Implementing the Party's Policies Regarding TCM and Cultivating TCM Practitioners," and lent great support in areas of human resources, finance, and supplies, vigorously promoting the development of TCM. It is stipulated in the Constitution of the PRC that the state promotes modern medicine and traditional Chinese medicine to protect the people's health. In 1986, the State Council set up a relatively independent administration of TCM.

各省、自治区、直辖市也相继成立了中医药管理机构，为中医药发展提供了组织保障。第七届全国人民代表大会第四次会议将"中西医并重"列为新时期中国卫生工作五大方针之一。2003年，国务院颁布实施《中华人民共和国中医药条例》，2009年，国务院颁布实施《关于扶持和促进中医药事业发展的若干意见》，逐步形成了相对完善的中医药政策体系。

All provinces, autonomous regions, and municipalities directly under the central government have established their respective TCM administrations, which has laid an organizational basis for TCM development. At the Fourth Meeting of the Seventh National People's Congress, equal emphasis was put on Chinese and Western medicine, which was made one of the five guidelines in China's health work in the new period. In 2003 and 2009, the State Council issued the "Regulations of the People's Republic of China on Traditional Chinese Medicine" and the "Opinions on Supporting and Promoting the Development of Traditional Chinese Medicine," gradually forming a relatively complete policy system on TCM.

中国共产党第十八次全国代表大会以来，党和政府把发展中医药摆上更加重要的位置，作出一系列重大决策部署。在全国卫生与健康大会上，习近平总书记强调，要"着力推动中医药振兴发展"。中国共产党第十八次全国代表大会和十八届五中全会提出"坚持中西医并重""扶持中医药和民族医药事业发展"。

Since the CPC's 18th National Congress in 2012, the Party and the government have granted greater importance to the development of TCM, and made a series of major policy decisions and adopted a number of plans in this regard. At the National Conference on Hygiene and Health held in August 2016, President Xi Jinping emphasized the importance of revitalizing and developing traditional Chinese medicine. The CPC's 18th National Congress and the Fifth Plenary Session of the 18th CPC Central Committee both reiterated the necessity to pay equal attention to the development of traditional Chinese medicine and Western medicine and lend support to the development of TCM and ethnic minority medicine.

2015年，国务院常务会议通过《中医药法（草案）》，并提请全国人大常委会审议，为中医药事业发展提供良好的政策环境和法制保障。2016年，中共中央、国务院印发《"健康中国2030"规划纲要》，作为今后15年推进健康中国建设的行动纲领，提出了一系列振兴中医药发展、服务健康中国建设的任务和举措。

In 2015, the executive meeting of the State Council approved the "Law on Traditional Chinese Medicine (draft)" and submitted it to the Standing Committee of the National People's Congress for deliberation and approval, intending to provide a sounder policy environment and legal basis for TCM development. In 2016 the CPC Central Committee and the State Council issued the "Outline of the Healthy China 2030 Plan", a guide to improving the health of the Chinese people in the coming 15 years. It sets out a series of tasks and measures to implement the program and develop TCM.

国务院印发《中医药发展战略规划纲要（2016—2030年）》，把中医药发展上升为国家战略，对新时期推进中医药事业发展作出系统部署。这些决策部署，描绘了全面振兴中医药、加快医药卫生体制改革、构建中国特色医药卫生体系、推进健康中国建设的宏伟蓝图，中医药事业进入新的历史发展时期。

The State Council issued the "Outline of the Strategic Plan on the Development of Traditional Chinese Medicine (2016—2030)", which made TCM development a national strategy, with systemic plans for TCM development in the new era. These decisions and plans have mapped out a grand blueprint that focuses on the full revitalization of TCM, accelerated reform of the medical and healthcare system, the building of a medical and healthcare system with Chinese characteristics, and the advancement of the healthy China plan, thus ushering in a new era of development for TCM.

中国发展中医药的基本原则和主要措施
The basic principles and main measures envisioned to develop TCM

坚持以人为本，实现中医药成果人民共享。中医药有很深的群众基础，文化理念易于为人民群众所接受。中医药工作以满足人民群众健康需求为出发点和落脚点，不断扩大中医医疗服务供给，提高基层中医药健康管理水平，推进中医药与社区服务、养老、旅游等融合发展，普及中医药健康知识，倡导健康的生产生活方式，增进人民群众健康福祉，保证人民群众享有安全、有效、方便的中医药服务。

Putting people first, and making achievements in TCM development accessible to everyone. TCM roots deep among the public, and the philosophies it contains are easy to understand. To meet the people's demand for healthcare, China endeavors to expand the supply of TCM services, improve community-level TCM health management, advance the integral development of TCM with community service, care of the elderly and tourism, spread knowledge of TCM and advocate healthy ways of life and work, enhance welfare for the public, and ensure that the people can enjoy safe, efficient, and convenient TCM services.

坚持中西医并重，把中医药与西医药摆在同等重要的位置。坚持中医药与西医药在思想认识、法律地位、学术发展和实践应用上的平等地位，健全管理体制，加大财政投入，制定体现中医药自身特点的政策和法规体系，促进中、西医药协调发展，共同为维护和增进人民群众健康服务。

Equal attention to TCM and Western medicine. Equal status shall be accorded to TCM and Western medicine in terms of ideological understanding, legal status, academic development, and practical application. Efforts shall be made to improve system of administration related to TCM, increase financial input, formulate policies, laws and regulations suited to the unique features of TCM, promote coordinated development of TCM and Western medicine, and make sure that they both serve the maintenance and improvement of the people's health.

坚持中医与西医相互取长补短、发挥各自优势。坚持中西医相互学习，组织西医学习中医，在中医药高等院校开设现代医学课程，加强高层次中西医结合人才培养。中医医院在完善基本功能基础上，突出特色专科专病建设，推动综合医院、基层医疗卫生机构设置中医药科室，实施基本公共卫生服务中医药项目，促进中医药在基本医疗卫生服务中发挥重要作用。建立健全中医药参与突发公共事件医疗救治和重大传染病防治的机制，发挥中医药独特优势。

Making TCM and Western medicine complementary to each other, and letting each play to its strengths. The state encourages exchanges between TCM and Western medicine, and creates opportunities for Western medical practitioners to learn from their TCM counterparts. Modern medicine courses are offered at TCM colleges and universities to strengthen the cultivation of doctors who have a good knowledge of both TCM and Western medicine. In addition to the general departments, TCM hospitals have been encouraged to open specialized departments for specific diseases. General hospitals and community-level medical care organizations have been encouraged to set up TCM departments, and TCM has been made available to patients in the basic medical care system and efforts have been made to make it play a more important role in basic medical care. A mechanism has been established for TCM to participate in medical relief of public emergencies and the prevention and control of serious infectious diseases.

坚持继承与创新的辩证统一，既保持特色优势又积极利用现代科学技术。建立名老中医药专家学术思想和临床诊疗经验传承制度，系统挖掘整理中医古典医籍与民间医药知识和技术。建设符合中医药特点的科技创新体系，开展中医药基础理论、诊疗技术、疗效评价等系统研究，组织重大疑难疾病、重大传染病防治的联合攻关和对常见病、多发病、慢性病的中医药防治研究，推动中药新药和中医诊疗仪器、设备研制开发。

Upholding the dialectical unity of tradition and innovation, i.e., maintaining TCM's characteristics while actively applying modern science and technology in TCM development. A system has been established to carry forward the theories and clinical experience of well-known veteran TCM experts, and efforts have been made to rediscover and categorize ancient TCM classics and folk medical experience and practices. A system of technological innovation has been established to advance TCM progress, and efforts have been made to carry out systemic research on the fundamental theories, clinical diagnosis and treatment, and therapeutic evaluation of TCM. Interdisciplinary efforts have been organized in joint research on the treatment and control of major difficult and complicated diseases and major infectious diseases, as well as research on the prevention and treatment of common diseases, frequently occurring diseases, and chronic diseases using TCM. Endeavor has been made in the R&D of new TCM medicines, and medical devices and equipment.

坚持统筹兼顾，推进中医药全面协调可持续发展。把中医药医疗、保健、科研、教育、产业、文化作为一个有机整体，统筹规划、协调发展。实施基层服务能力提升工程，健全中医医疗服务体系。实施"治未病"健康工程，发展中医药健康服务。开展国家中医临床研究基地建设，构建中医药防治重大疾病协同创新体系。实施中医药传承与创新人才工程，提升中医药人才队伍素质。推动中药全产业链绿色发展，大力发展非药物疗法。推动中医药产业升级，培育战略性新兴产业。开展"中医中药中国行"活动，弘扬中医药核心价值理念。

Making overall plans for integrated, coordinated and sustainable development of TCM. The state makes overall plans for the coordinated development of TCM, integrating such areas as TCM clinical practices, healthcare, R&D, education, industry, and culture. Efforts have been made to improve community-level TCM services and the TCM medical care system. A health promotion project featuring preventative treatment of diseases has been launched to enhance TCM medical care. China has built research bases for TCM clinical studies, developed a system of coordination and innovation for the prevention and treatment of major diseases with TCM, and launched programs for training professionals necessary for TCM inheritance and innovation, and improving the quality of the ranks of TCM workers. The state has set out to promote the green development of the entire TCM industrial chain, and vigorous efforts have been made in the development of non-pharmacological therapies. Further efforts have been made to upgrade the TCM industry and develop it into an emerging strategic industry. A nationwide program has been conducted to promote the core values of TCM.

坚持政府扶持、各方参与，共同促进中医药事业发展。把中医药作为经济社会发展的重要内容，纳入相关规划、给予资金支持。强化中医药监督管理，实施中医执业医师、医疗机构和中成药准入制度，健全中医药服务和质量安全标准体系。制定优惠政策，充分发挥市场在资源配置中的决定性作用，积极营造平等参与、公平竞争的市场环境，不断激发中医药发展的潜力和活力。鼓励社会捐资支持中医药事业，推动社会力量开办中医药服务机构。

Promoting TCM development by way of government support and multi-party participation. The government has made TCM an important component of economic and social development, and has included it in relevant plans and provided financial support. The state has strengthened the supervision and administration of TCM practices, and initiated a market access system for TCM practitioners, TCM medical institutions, and TCM medicines, and improved the standards for the quality and safety of TCM service. The government has developed preferential policies to let the market play a full and decisive role in allocating resources, and is striving to create a market environment characterized by equal participation and fair play, so as to maximize the potential and vitality of TCM. Encouragement has been given to social capital to support the development of TCM, and to private investors to establish TCM healthcare institutions.

三、中医药的传承与发展
III. Carrying Forward the Tradition and Ensuring the Development of TCM

内容参见二维码

生词和词组
New Words and Expressions

中医药	Traditional Chinese Medicine (TCM)
远古时代	in remote antiquity
病痛	ailments and pains
食医	dietician
疾医	physician
疡医	doctor of decoctions
兽医	veterinarian
治未病	preventive treatment of disease
病理	pathology
经络	meridians & collaterals
针灸	acupuncture & moxibustion
大医精诚	mastership of medicine lies in proficient medical skills and lofty medical ethics
推拿	tuina (massage)
拔罐	cupping
刮痧	guasha (spooning)
小夹板	the small splints used in Chinese osteopathy

天花	smallpox
种痘	vaccination
青蒿素	artemisinin
乳香	frankincense
没药	myrrh
国务院常务会议	the executive meeting of the State Council
全国人大常委会	the Standing Committee of the National People's Congress
审议	deliberation and approval
文化理念	philosophies
辩证统一	dialectical unity
传染性非典型肺炎	SARS (severe acute respiratory syndrome)
甲型 H1N1 流感	influenza A virus subtype H1N1
艾滋病	HIV/AIDS (Human Immunodeficiency Virus and Acquired Immune Deficiency Syndrome)
手足口病	HFMD (Hand, Foot and Mouth Disease)
人感染 H7N9 禽流感	Influenza A virus subtype H7N9 in humans
传染病	infectious diseases
慢性非传染病	chronic non-infectious diseases
中药砷剂 TCM	compound arsenic trioxide
急性早幼粒细胞白血病	acute promyelocytic leukemia (APL)
野生抚育	wild tending
中医药理论	theories of traditional Chinese medicine and pharmacology
丸、散、膏、丹	pills, powders, ointments and pellets
滴丸	dropping pills
片剂	tablets
膜剂	pods
胶囊	capsules
疑难重症	complicated and refractory diseases

翻译探讨
Translation study and Discussion

中文：随着人们健康观念变化和医学模式转变，中医药越来越显示出其独特价值。

英文：As ideas on fitness and medical models change and evolve, traditional Chinese medicine has come to underline a more and more profound value.

提示："显示"常常翻译成"show, indicate, demonstrate"。这里的"come to underline"值得借鉴并适当应用。

中文：新中国成立以来，中国高度重视和大力支持中医药发展。

英文：Since the founding of the People's Republic of China in 1949, the Chinese government has set great store by TCM and rendered vigorous support to its development.

提示："重视"往往翻译成"attach (great) importance to, pay (great) attention to, take…seriously"。这里的"set great store"的使用同样显示出对原文正式、庄严文体的语言之转换。"重视"英文也可以是"lay great store"，如后文"中国高度重视中医药事业发展"英文为"China lays great store by the development of TCM"。

中文：春秋战国（前770—前221）时期，扁鹊总结前人经验，提出"望、闻、问、切"四诊合参的方法，奠定了中医临床诊断和治疗的基础。

英文：During the Spring and Autumn and Warring States Period (770—221 BC), Bian Que drew on the experience of his predecessors and put forward the four diagnostic methods-inspection, auscultation & olfaction, inquiry, and palpation, laying the foundation for TCM diagnosis and treatment.

提示："望、闻、问、切"翻译成"inspection, auscultation & olfaction, inquiry, and palpation"，注意每个词，尤其是"闻"的翻译"auscultation & olfaction"。

中文：……形成了中医药理论体系框架。
英文：A theoretical framework for TCM had been in place.
提示：注意译文主语的选择。

中文：……一些学者开始探索中西医药学汇通、融合。
英文：some TCM experts began to explore ways to absorb the essence of Western medicine for a combination of TCM with Western medicine.
提示：注意英文的补充如"TCM"；注意"汇通、融合"的翻译处理，该句"融合"译作"for a combination of …"。

中文：在数千年的发展过程中，中医药不断吸收和融合各个时期先进的科学技术和人文思想……
英文：During its course of development spanning a couple of millennia, TCM has kept drawing and assimilating advanced elements of natural science and humanities……
提示：注意此句"融合"的翻译处理。

中文：第二，注重"平"与"和"。
英文：Second, setting great store by the principle of harmony.
提示：此句可作术语之意译。

中文：维护健康就是维护人的整体功能动态平衡……
英文：Therefore, maintaining health actually means conserving the dynamic balance of body functions……
提示：注意上下文逻辑关系分析和适当添加表示逻辑关系的词语"Therefore"；注意中文两个"维护"的不同英文翻译。

中文：中医诊疗着眼于"病的人"而不仅是"人的病"，着眼于调整致病因子作用于人体后整体功能失调的状态。
英文：TCM therapies focus on the person who is sick rather than the illness that the patient contracts, i. e., aiming to restore the harmonious state of body functions that is disrupted by pathogenic factors.

提示：此处中文"病的人"和"人的病"为"交错排列法（chiasmus）"，英文则为意译。

中文：中医强调生活方式和健康有着密切关系，主张以养生为要务……

英文：TCM believes that lifestyle is closely related to health, so it advocates health should be preserved in daily life.

提示：此处"养生"英文为"health should be preserved in daily life"。

中文：中医药是中华优秀传统文化的重要组成部分和典型代表，强调"道法自然、天人合一""阴阳平衡、调和致中""以人为本、悬壶济世"，体现了中华文化的内核。

英文：TCM is an important component and a characteristic feature of traditional Chinese culture. Applying such principles as "man should observe the law of the nature and seek for the unity of the heaven and humanity," "yin and yang should be balanced to obtain the golden mean," and "practice of medicine should aim to help people," TCM embodies the core value of Chinese civilization.

提示：中文一大特点即运用四字格。该句段中四字格大量使用。英文则采取意译方式，必要时还可以省译，以求"舍形取义"，即不受原文字面结构束缚，译取原文的意义。

中文：同时，乳香、没药等南药的广泛引进，丰富了中医药的治疗手段。

英文：Meanwhile, massive imports of medicinal substances such as frankincense and myrrh have enriched TCM therapies.

提示："therapy"作不可数名词用时为"治疗"或"疗法"，作可数名词尤其是复数名词时则抽象概念具体化，意为"各种治疗手段"。

中文：1978年，中共中央转发卫生部《关于认真贯彻党的中医政策，解决中医队伍后继乏人问题的报告》，并在人、财、物等方面给予大力支持，有力地推动了中医药事业发展。

英文：In 1978, the Communist Party of China (CPC) Central Committee transmitted throughout the country the Ministry of Health's "Report on Implementing the Party's Policies Regarding TCM and Cultivating TCM Practitioners," and lent great support in areas of human resources, finance, and supplies, vigorously promoting the development of TCM.

提示：《报告》内容的英文，和原文汉语对比可视作正反倒说的翻译技巧。"支持"的英文短语动词为"lend"，整个短语"大力支持"原形为"lend great support"。

中文：中华人民共和国宪法指出，发展现代医药和我国传统医药，保护人民健康。

英文：It is stipulated in the Constitution of the PRC that the state promotes modern medicine and traditional Chinese medicine to protect the people's health.

提示：这里的"指出"实际是"规定"。翻译离不开精密透彻分析，须还原词语在上下文中的确切意思。

中文：……国务院颁布实施《中华人民共和国中医药条例》……

英文：… the State Council issued the "Regulations of the People's Republic of China on Traditional Chinese Medicine" …

提示："颁布实施"英文为一个词语"issued"；下文"印发"为同一词语。

中文：中国共产党第十八次全国代表大会以来，党和政府把发展中医药摆上更加重要的位置，作出一系列重大决策部署。

英文：Since the CPC's 18th National Congress in 2012, the Party and the government have granted greater importance to the development of TCM, and made a series of major policy decisions and adopted a number of plans in this regard.

提示："摆上更加重要的位置"英文为"have granted greater importance to…"；"作出一系列重大决策部署"英文为"(have) made a series of major policy decisions and adopted a number of plans".

中文：2015年，国务院常务会议通过《中医药法（草案）》，并提请全国人大常委会审议，为中医药事业发展提供良好的政策环境和法制保障。

英文：In 2015, the executive meeting of the State Council approved the "Law on Traditional Chinese Medicine（draft）" and submitted it to the Standing Committee of the National People's Congress for deliberation and approval, intending to provide a sounder policy environment and legal basis for TCM development.

提示：注意"良好的"一词的英文。

中文：坚持以人为本，实现中医药成果人民共享。

英文：Putting people first, and making achievements in TCM development accessible to everyone.

提示：注意此处"共享"的英文表达"make…accessible to everyone"。

中文：中医药工作以满足人民群众健康需求为出发点和落脚点……

英文：To meet the people's demand for healthcare…

提示：此句为省译式意译。

中文：坚持中西医并重……

英文：Equal attention to TCM and Western medicine

提示：此为标题式翻译，同时将原文动宾结构名词化，和全文相应部分并列。

中文：……把中医药与西医药摆在同等重要的位置……

英文：Equal status shall be accorded to TCM and Western medicine……

提示：这里把字句进行被动翻译处理，同时此句与以下部分进行合译处理。

中文：强化中医药监督管理，实施中医执业医师、医疗机构和中成药准入制度，健全中医药服务和质量安全标准体系。

英文：The state has strengthened the supervision and administration of TCM practices, and initiated a market access system for TCM practitioners, TCM medical institutions, and TCM medicines, and improved the standards for the quality and safety of TCM service.

提示：中文原文为无主句，英文添加逻辑主语。下句情况类似，英译添加主语为"The government"。

中文：中医药健康管理项目作为单独一类列入国家基本公共卫生服务项目，中医药在公共卫生服务中的潜力和优势正逐步释放，推动卫生发展模式从重疾病治疗向全面健康管理转变。

英文：TCM health management program has been incorporated as a separate category into the national basic public health service program, gradually releasing the potential and strengths of TCM in public health services, thus fostering a shift from treatment of serious diseases to comprehensive health management in the mode of health development.

提示：汉语的连动式并列结构转译为英语的主从结构，英文谓语动词为"has been incorporated"，"releasing…"和"fostering…"为非谓语动词结构作状语。

中文：实施中医药传承与创新人才工程，开展第五批全国名老中医药专家学术经验继承工作，建设了1016个全国名老中医药专家传承工作室、200个全国基层名老中医药专家传承工作室，为64个中医学术流派建立传承工作室。

英文：In the course of implementing the program of training professionals for inheritance and innovation in TCM, efforts have been made to conserve and disseminate the academic ideas and practical experience of the fifth batch of prominent TCM experts. By 2015, 1,016 studios had been set up for carrying forward their expertise; 200 studios had been set up for passing on the expertise of prominent TCM experts at the grassroots level; 64 studios had been established for promoting various schools of TCM.

提示：中文原文有四个"传承"，英文里用不同的词语，包括"inheritance"，"carrying forword（their expertise）"，"passing on（the expertise of…）"，"promoting"，为"一词多译"。同时"继承"英文为"conserve and disseminate"。

中文：因将传统中药的砷剂与西药结合治疗急性早幼粒细胞白血病的疗效明显提高，王振义、陈竺获得第七届圣捷尔吉癌症研究创新成就奖。

英文：Wang Zhenyi and Chen Zhu were awarded the Seventh Annual Szent-Gyorgyi Prize for Progress in Cancer Research for combining the Western medicine ATRA and the TCM compound arsenic trioxide to treat acute promyelocytic leukemia（APL）.

提示：中文原因在前，英译为倒装语序，即先（成果）结果后原因。该句英文结构也因此和前句一致。

中文：中药已从丸、散、膏、丹等传统剂型，发展到现在的滴丸、片剂、膜剂、胶囊等40多种剂型，中药产品生产工艺水平有了很大提高，基本建立了以药材生产为基础、工业为主体、商业为纽带的现代中药产业体系。

英文：The dosage forms of TCM medicines have increased from a traditionally limited number of forms such as pills, powders, ointments and pellets into more than 40, including dropping pills, tablets, pods and capsules, indicating marked improvement in the technological level of Chinese medicinal drug production, and initial establishment of a modern Chinese medicine industry based on the production of medicinal materials and industrial production and tied together by commerce.

提示：注意专有名词"丸、散、膏、丹"，"滴丸、片剂、膜剂、胶囊"等的翻译；"很大提高"的"很大"不要按字面译为"very big"，这里英文是"marked"，意为"great"，"considerable"，"significant"，"remarkable"等。

第一部分　农业政策 | 041

中文：2015年中药工业总产值7866亿元，占医药产业规模的28.55%，成为新的经济增长点……

英文：In 2015, the total output value of the TCM pharmaceutical industry was RMB786.6 billion, accounting for 28.55 percent of the total generated by the country's pharmaceutical industry...

提示：注意中英文数字的转换和百分比数字的翻译。

中文：加强中医药健康知识的宣传普及，持续开展"中医中药中国行"大型科普活动，利用各种媒介和中医药文化宣传教育基地，向公众讲授中医药养生保健、防病治病的基本知识和技能，全社会利用中医药进行自我保健的意识和能力不断增强，促进了公众健康素养的提高。

英文：Efforts have been reinforced to promote public awareness in TCM healthcare, including events under the campaign of Traditional Chinese Medicine Across China. Public talks have been organized through media and TCM education bases popularizing basic knowledge and skills of TCM healthcare and prevention and treatment of illnesses. In this way, public awareness of and ability to practice TCM healthcare has been enhanced, and general public health has improved.

提示：整个段落为主动转被动（语态）翻译。

中文：中医药已成为中国与东盟、欧盟、非洲、中东欧等地区和组织卫生经贸合作的重要内容，成为中国与世界各国开展人文交流、促进东西方文明交流互鉴的重要内容，成为中国与各国共同维护世界和平、增进人类福祉、建设人类命运共同体的重要载体。

英文：Traditional Chinese medicine has become an important area of health and trade cooperation between China and the ASEAN, EU, Africa, and Central and Eastern Europe, a key component in people-to-people exchanges between China and the rest of the world and between Eastern and Western cultures, and an important vehicle for China and other countries to work together in promoting world peace, improving the well-being of humanity, and developing a community of shared future.

提示：中文原文有三个"成为"，为排比句，英文则只用一次动词谓语"has become",并使用同位语结构。两个"重要内容"则分别译为"an important area"和"a key component",避免重复。

中文:积极推动传统药监督管理国际交流与合作,保障传统药安全有效。

英文:China is working actively to promote international exchange and cooperation in the supervision and management of traditional medicine, in an effort to ensure that it is safe and effective.

提示:中文为无主句,连动式并列结构。英文添加主语和相关逻辑连接。

练习
Exercises

练习一　将课文原文内容做英汉互译练习

练习二　阅读理解

In this section there are passages followed by questions or unfinished statements, each with four suggested answers marked A, B, C, and D. Choose the one that you think is the best answer.

An upsurge of new research suggests animals have a much higher level of brainpower than previously thought. Before defining animals' intelligence, scientists defined what is not intelligence. Instinct is not intelligence. It is a skill programmed into an animal's brain by its genetic heritage. Rote conditioning or cuing, in which animals learn to do or not to do certain things by following outside signals is also not intelligence, since tricks can be learned by repetition, but no real thinking is involved. Scientists believe insight, the ability to use tools, and communications using human language are effective measures.

When judging animal intelligence, scientists look for insight, which they define as a flash of sudden understanding. When a young gorilla could not reach fruit from a tree, she noticed crates scattered about the lawn, piled them and then climbed on them to reach her reward. The gorilla's insight allowed her to solve a new problem without trial and error. The ability to use tools is also an important sign of intelligence. Crows use sticks to pry peanuts out of cracks. The crow exhibits intelligence by showing it has learned what a stick can do. Likewise, otters use rocks to crack open crab and, in a series of complex moves, chimpanzees have been known to use sticks to get at their favorite snack-termites. Many animals have learned to communicate using human language. Some primates have learned hundreds of words in sign language. One chimp can recognize and correctly use more than 250 abstract symbols on a keyboard and one parrot can distinguish five objects of two different types and can understand the difference between numbers, colors, and kinds of object.

The research on animal intelligence raises important questions. If animals are smarter than once thought, would that change the way humans interact with them? Would humans stop hunting them for sport or survival? Would animals still be used for food or clothing or medical experimentation? Finding the answer to these tough questions makes a difficult puzzle even for a large-brained, problem-solving species like our own.

1. According to the text, which is true about animals communicating through the use of human language?

   A. Parrots can imitate or repeat a sound.

   B. Dolphins click and whistle.

   C. Crows screech warnings to other crows.

   D. Chimps have been trained to use sign language or word symbolizing geometric shapes.

2. The underlined word "upsurge", (line 1, para 1), most nearly means _____.

   A. an increasingly large amount

   B. a decreasing amount

C. a well-known amount

D. an immeasurable amount

3. The chimpanzee's ability to use a tool illustrates high intelligence because _____.

A. he is able to get his food and eat it

B. he faced a difficult task and accomplished it

C. he stored knowledge away and called it up at the right time

D. termites are protein-packed

4. It can be inferred from the concluding paragraph of this text that _____.

A. there is no definitive line between those animals with intelligence and those without

B. animals are being given opportunities to display their intelligence

C. research showing higher animal intelligence may fuel debate on ethics and cruelty.

D. animals are capable of untrained thought well beyond mere instinct

5. Which of the following is NOT a sign of animal intelligence?

A. Shows insight.　　　　　　B. Cues.

C. Uses tools.　　　　　　　　D. Makes a plan.

第三单元 2018年政府工作报告（节选）

——2019年3月5日在第十三届全国人民代表大会第二次会议上

国务院总理 李克强

UNIT THREE
REPORT ON THE WORK OF THE GOVERNMENT IN 2018（EXCERPTED）

Delivered at the Second Session of the 13th National People's Congress of the People's Republic of China on March 5，2019

Premier of the State Council

Li Keqiang

各位代表：

Fellow Deputies，

现在，我代表国务院，向大会报告政府工作，请予审议，并请全国政协委员提出意见。

On behalf of the State Council，I will now report to you on the work of the government and ask for your deliberation and approval. I also invite comments from members of the National Committee of the Chinese People's Political Consultative Conference（CPPCC）.

一、2018年工作回顾
Ⅰ. 2018 in Review

过去一年是全面贯彻党的十九大精神开局之年，是本届政府依法履职第一年。我国发展面临多年少有的国内外复杂严峻形势，经济出现新的下行压力。在以习近平同志为核心的党中央坚强领导下，全国各族人民以习近平新时代中国特色社会主义思想为指导，砥砺奋进，攻坚克难，完成全年经济社会发展主要目标任务，决胜全面建成小康社会又取得新的重大进展。

The year 2018 was the first year for putting the guiding principles of the 19th National Congress of the Communist Party of China fully into effect. It was also this government's first to perform, in accordance with law, the functions of office. In pursuing development this year, China faced a complicated and challenging domestic and international environment of a kind rarely seen in many years, and its economy came under new downward pressure. Under the firm leadership of the Party Central Committee with Comrade Xi Jinping at its core, we, the Chinese people of all ethnic groups, guided by Xi Jinping Thought on Socialism with Chinese Characteristics for a New Era, forged ahead and overcame difficulties. The year's main targets for economic and social development were accomplished, and in building a moderately prosperous society in all respects, we made major progress toward a decisive victory.

——经济运行保持在合理区间。国内生产总值增长6.6%,总量突破90万亿元。经济增速与用电、货运等实物量指标相匹配。居民消费价格上涨2.1%。国际收支基本平衡。城镇新增就业1361万人、调查失业率稳定在5%左右的较低水平。近14亿人口的发展中大国,实现了比较充分就业。

The main economic indicators were kept within an appropriate range. Gross domestic product (GDP) grew by 6.6 percent, exceeding 90 trillion yuan. Economic growth matched electricity consumption, freight transport, and other indicators. Consumer prices rose by 2.1 percent. In the balance of payments a basic equilibrium was maintained. A further 13.61 million new urban jobs were added, and the surveyed unemployment rate remained stable at a comparatively low level of around 5 percent. A big developing country with a population close to 1.4 billion like ours has achieved relatively full employment.

——经济结构不断优化。消费拉动经济增长作用进一步增强。服务业对经济增长贡献率接近60%,高技术产业、装备制造业增速明显快于一般工业,农业再获丰收。单位国内生产总值能耗下降3.1%。质量和效益继续提升。

Economic structure was further improved. Consumption continued to play an increasing role in driving economic growth. The service sector's contribution to growth approached 60 percent. Growth in high-tech industries and equipment manufacturing outstripped that of other industries. Harvests were again good. Energy consumption per unit of GDP fell by 3.1 percent. The quality and returns of growth continued to improve.

——发展新动能快速成长。嫦娥四号等一批重大科技创新成果相继问世。新兴产业蓬勃发展，传统产业加快转型升级。大众创业、万众创新深入推进，日均新设企业超过1.8万户，市场主体总量超过1亿户。新动能正在深刻改变生产生活方式、塑造中国发展新优势。

New growth drivers grew rapidly. A number of major scientific and technological innovations were made, like the Chang'e-4 lunar probe. Emerging industries thrived and traditional industries saw faster transformation and upgrading. Business startups and innovation continued to surge nationwide, with an average of over 18,000 new businesses opening daily and the total number of market entities passing the 100 million mark. New growth drivers are now profoundly changing our mode of production and way of life, creating new strengths for China's development.

——改革开放取得新突破。国务院及地方政府机构改革顺利实施。重点领域改革迈出新的步伐，市场准入负面清单制度全面实行，简政放权、放管结合、优化服务改革力度加大，营商环境国际排名大幅上升。对外开放全方位扩大，共建"一带一路"取得重要进展。首届中国国际进口博览会成功举办，海南自贸试验区启动建设。货物进出口总额超过30万亿元，实际使用外资1383亿美元、稳居发展中国家首位。

New breakthroughs were made in reform and opening up. Institutional reforms of both the State Council and local governments were implemented smoothly. New progress was made in reform in key fields. The negative list system for market access was put fully into effect. Reforms to streamline administration and delegate power, improve regulation, and upgrade services were intensified, and our business environment rose significantly in international rankings. Opening up was expanded on all fronts, and joint efforts to pursue the Belt and Road Initiative (BRI) made significant headway. The first China International Import Expo was a success. Work began on building the China (Hainan) Pilot Free Trade Zone. China's total volume of trade in goods exceeded 30 trillion yuan, and its utilized foreign investment totaled US＄138.3 billion, ranking China first among developing countries.

——三大攻坚战开局良好。防范化解重大风险，宏观杠杆率趋于稳定，金融运行总体平稳。精准脱贫有力推进，农村贫困人口减少1386万，易地扶贫搬迁280万人。污染防治得到加强，细颗粒物（PM2.5）浓度继续下降，生态文明建设成效显著。

The three critical battles got off to a good start. We forestalled and defused major risks. The macro leverage ratio trended toward a stable level; the financial sector was generally stable. Precision poverty alleviation made significant progress, with the rural poor population reduced by 13.86 million, including 2.8 million people assisted through relocation from inhospitable areas. Pollution prevention and control was strengthened, and PM2.5 density continued to fall. Marked achievements were made in ecological conservation.

——人民生活持续改善。居民人均可支配收入实际增长6.5％。提高个人所得税起征点，设立6项专项附加扣除。加大基本养老、基本医疗等保障力度，资助各类学校家庭困难学生近1亿人次。棚户区住房改造620多万套，农村危房改造190万户。城乡居民生活水平又有新的提高。

Living standards continued to improve. Per capita disposable personal income grew by 6.5 percent in real terms. The threshold for individual income tax was raised and six special additional deductions were created. Support for basic elderly care and basic health care was strengthened. Close to 100 million payments were made to assist students from families in financial difficulty, covering all school types. More than 6.2 million housing units were rebuilt in rundown urban areas and 1.9 million dilapidated rural houses were renovated. Urban and rural living standards continued to rise.

我们隆重庆祝改革开放40周年，深刻总结改革开放的伟大成就和宝贵经验，郑重宣示在新时代将改革开放进行到底的坚定决心，激励全国各族人民接续奋斗，再创新的历史伟业。

We solemnly commemorated the 40th anniversary of reform and opening up, thoroughly reviewed its great achievements and the valuable experience gained in its pursuit, and pledged our resolve to see reform and opening up through in the new era, thus galvanizing the Chinese people of all ethnic groups to continue their hard work to make new historic achievements.

回顾过去一年，成绩来之不易。我们面对的是深刻变化的外部环境。经济全球化遭遇波折，多边主义受到冲击，国际金融市场震荡，特别是中美经贸摩擦给一些企业生产经营、市场预期带来不利影响。我们面对的是经济转型阵痛凸显的严峻挑战。新老矛盾交织，周期性、结构性问题叠加，经济运行稳中有变、变中有忧。我们面对的是两难多难问题增多的复杂局面。实现稳增长、防风险等多重目标，完成经济社会发展等多项任务，处理好当前与长远等多种关系，政策抉择和工作推进的难度明显加大。经过全国上下共同努力，我国经济发展在高基数上总体平稳、稳中有进，社会大局保持稳定。这再次表明，在中国共产党领导下，中国人民有战胜任何艰难险阻的勇气、智慧和力量，中国的发展没有过不去的坎。

Looking back at the past year, we can see that our achievements did not come easily. What we faced was profound change in our external environment. Setbacks in economic globalization, challenges to multilateralism, shocks in the international financial market, and especially the China-US economic and trade frictions, had an adverse effect on the production and business operations of some companies and on market expectations. What we faced were severe challenges caused by the growing pains of economic transformation. An interlacing of old and new issues and a combination of cyclical and structural problems brought changes in what was a generally stable economic performance, some of which caused concern. What we faced was a complicated terrain of increasing dilemmas. We had multiple targets to attain, like ensuring stable growth and preventing risks; multiple tasks to complete, like promoting economic and social development; and multiple relationships to handle, like that between short-term and long-term interests. And the difficulty of making policy choices and moving work forward increased markedly. With the concerted efforts of the whole country, the Chinese economy, from a larger base, achieved generally stable growth while making further progress; and social stability was ensured. This once again shows that the Chinese people, under the leadership of the Communist Party of China, have the courage, vision, and strength to prevail over any difficulty or obstacle. There is no difficulty that cannot be overcome in China's pursuit of development!

二、2019年经济社会发展总体要求和政策取向
Ⅱ. Economic and Social Development in 2019: Overall Requirements and Policy Directions

今年是新中国成立70周年,是全面建成小康社会、实现第一个百年奋斗目标的关键之年。做好政府工作,要在以习近平同志为核心的党中央坚强领导下,以习近平新时代中国特色社会主义思想为指导,全面贯彻党的十九大和十九届二中、三中全会精神,统筹推进"五位一体"总体布局,协调推进"四个全面"战略布局,坚持稳中求进工作总基调,坚持新发展理念,坚持推动高质量发展,坚持以供给侧结构性改革为主线,坚持深化市场化改革、扩大高水平开放,加快建设现代化经济体系,继续打好三大攻坚战,着力激发微观主体活力,创新和完善宏观调控,统筹推进稳增长、促改革、调结构、惠民生、防风险、保稳定工作,保持经济运行在合理区间,进一步稳就业、稳金融、稳外贸、稳外资、稳投资、稳预期,提振市场信心,增强人民群众获得感、幸福感、安全感,保持经济持续健康发展和社会大局稳定,为全面建成小康社会收官打下决定性基础,以优

异成绩庆祝中华人民共和国成立70周年。

This year is the 70th anniversary of the founding of the People's Republic of China. It will be a crucial year for us as we endeavor to achieve the first Centenary Goal of building a moderately prosperous society in all respects. To fulfill the work of government, under the strong leadership of the Party Central Committee with Comrade Xi Jinping at its core, we must:

• follow the guidance of Xi Jinping Thought on Socialism with Chinese Characteristics for a New Era;

• implement fully the guiding principles of the Party's 19th National Congress and the second and third plenary sessions of its 19th Central Committee;

• pursue coordinated progress in the five-sphere integrated plan;

• pursue balanced progress in the four-pronged comprehensive strategy;

• adhere to the general principle of pursuing progress while ensuring stability;

• continue to apply the new development philosophy;

• continue to work for high-quality development;

• continue to pursue supply-side structural reform as our main task;

• continue to deepen market-orientated reforms and expand high-standard opening up;

• work faster to modernize the economy;

• continue the three critical battles;

• invigorate micro entities;

• explore innovations in and improve macro regulation;

• make coordinated efforts to maintain stable growth, advance reform, make structural adjustments, improve living standards, guard against risks and ensure stability.

• keep major economic indicators within an appropriate range;

• take further steps to ensure stable employment, a stable financial

sector, stable foreign trade, stable foreign investment, stable domestic investment, and stable expectations;
- boost market confidence;
- enable people to feel more satisfied, happy, and secure;
- sustain healthy economic development and maintain social stability.

By doing the above, we will create the pivotal underpinning for completing the building of a moderately prosperous society and celebrate with outstanding accomplishments the 70th anniversary of the founding of the People's Republic of China.

综合分析国内外形势,今年我国发展面临的环境更复杂更严峻,可以预料和难以预料的风险挑战更多更大,要做好打硬仗的充分准备。困难不容低估,信心不可动摇,干劲不能松懈。我国发展仍处于重要战略机遇期,拥有足够的韧性、巨大的潜力和不断迸发的创新活力,人民群众追求美好生活的愿望十分强烈。我们有战胜各种困难挑战的坚定意志和能力,经济长期向好趋势没有也不会改变。

A full analysis of developments in and outside China shows that in pursuing development this year, we will face a graver and more complicated environment as well as risks and challenges, foreseeable and otherwise, that are greater in number and size. We must be fully prepared for a tough struggle. The difficulties we face must not be underestimated, our confidence must not be weakened, and the energy we bring to our work must not be allowed to wane. China is still in an important period of strategic opportunity for development and has ample resilience, enormous potential, and great creativity to unleash. The longing of our people for a better life is strong. We have the unshakable will and the ability needed to prevail over difficulties and challenges of any kind, and our economic fundamentals are sound and will remain sound over the long term.

今年经济社会发展的主要预期目标是:国内生产总值增长6%~6.5%;城镇新增就业1100万人以上,城镇调查失业率5.5%左右,城镇登记失业率4.5%以内;居民消费价格涨幅3%左右;国际收支基本平衡,进出口稳中提质;宏观杠杆率基本稳定,金融财政风险有效防控;农村贫困人口减少1000万以上，居民收入增长与经济增长基本同步；生态环境进一步改善，单位国内生产总值能耗下降3%左右，主要污染物排放量继续下降。上述主要预期目标，体现了推动高质量发展要求，符合我国发展实际，与全面建成小康社会目标相衔接，是积极稳妥的。实现这些目标，需要付出艰苦努力。

With the above in mind, the main projected targets for economic and social development this year are set as follows:

- GDP growth of 6~6.5 percent;
- Over 11 million new urban jobs, a surveyed urban unemployment rate of around 5.5 percent, and a registered urban unemployment rate within 4.5 percent;
- CPI increase of around 3 percent;
- A basic equilibrium in the balance of payments, and stable, better-structured imports and exports;
- A macro leverage ratio that is basically stable, and effective prevention and control of financial and fiscal risks;
- A reduction of over 10 million in the rural poor population;
- Personal income growth that is basically in step with economic growth;
- A further improvement in the environment;
- A drop of around 3 percent in energy consumption per unit of GDP;
- Continued reductions in the discharge of major pollutants.

The above projected targets are ambitious but realistic-they represent our aim of promoting high-quality development, are in keeping with the current realities of China's development, and are aligned with the goal of completing the building of a moderately prosperous society in all respects. But to realize these goals we need to redouble our efforts.

要正确把握宏观政策取向，继续实施积极的财政政策和稳健的货币政策，实施就业优先政策，加强政策协调配合，确保经济运行在合理区间，促进经济社会持续健康发展。

We will ensure that the right direction is set for the pursuit of our macro policies. We will continue to pursue a proactive fiscal policy and a prudent monetary policy, implement an employment-first policy, and strengthen the coordination between these policies to keep major economic indicators within an appropriate range and sustain healthy economic and social development.

积极的财政政策要加力提效。今年赤字率拟按 2.8% 安排，比去年预算高 0.2 个百分点；财政赤字 2.76 万亿元，其中中央财政赤字 1.83 万亿元，地方财政赤字 9300 亿元。适度提高赤字率，综合考虑了财政收支、专项债券发行等因素，也考虑为应对今后可能出现的风险留出政策空间。今年财政支出超过 23 万亿元，增长 6.5%。中央对地方均衡性转移支付增长 10.9%。改革完善县级基本财力保障机制，缓解困难地区财政运转压力，决不让基本民生保障出问题。

We will pursue a proactive fiscal policy with greater intensity and enhance its performance. Deficit-to-GDP ratio this year is projected at 2.8 percent, a 0.2-percentage-point increase over that of last year. The budgetary deficit is projected at 2.76 trillion yuan, with a central government deficit of 1.83 trillion yuan and a local government deficit of 0.93 trillion yuan. In moderately increasing the deficit-to-GDP ratio, we have given full consideration to factors such as government revenue and expenditure and the issuance of special bonds; we have also taken into account the need to leave policy space to address risks that could arise in the future. Government expenditure is budgeted at over 23 trillion yuan, a 6.5 percent increase. The central government's transfer payments to local governments for equalizing access to basic public services will increase by 10.9 percent. We will reform and improve the mechanism for ensuring basic fiscal capacity at the county level, ease the pressure of budgetary constraints faced by localities, and make certain that people's basic living needs are met.

稳健的货币政策要松紧适度。广义货币 M2 和社会融资规模增速要与国内生产总值名义增速相匹配，以更好满足经济运行保持在合理区间的需要。在实际执行中，既要把好货币供给总闸门，不搞"大水漫灌"，又要灵活运用多种货币政策工具，疏通货币政策传导渠道，保持流动性合理充裕，有效缓解实体经济特别是民营和小微企业融资难融资贵的问题，防范化解金融风险。深化利率市场化改革，降低实际利率水平。完善汇率形成机制，保持人民币汇率在合理均衡水平上的基本稳定。

Our prudent monetary policy will be eased or tightened to the right degree. Increases in M2 money supply and aggregate financing should be in keeping with nominal GDP growth to keep major indicators within an appropriate range. In implementation, we will ensure the valve on aggregate monetary supply is well controlled and refrain from using a deluge of stimulus policies; but will also use flexibly a variety of monetary policy instruments to improve the transmission mechanism of monetary policy, maintain reasonably sufficient liquidity, effectively mitigate difficulties faced in the real economy, especially by private enterprises and small and micro businesses, in accessing affordable financing, and forestall and defuse financial risks. We will deepen reforms to strengthen the market's role in setting interest rates and lower real interest rates. We will improve the exchange rate mechanism and keep the RMB exchange rate generally stable and at an adaptive and balanced level.

就业优先政策要全面发力。就业是民生之本、财富之源。今年首次将就业优先政策置于宏观政策层面，旨在强化各方面重视就业、支持就业的导向。当前和今后一个时期，我国就业总量压力不减、结构性矛盾凸显，新的影响因素还在增加，必须把就业摆在更加突出位置。稳增长首要是为保就业。今年城镇新增就业要在实现预期目标的基础上，力争达到近几年的实际规模，既保障城镇劳动力就业，也为农业富余劳动力转移就业留出空间。只要就业稳、收入增，我们就更有底气。

An employment-first policy will be pursued with full force. Employment is the cornerstone of well-being, and the wellspring of wealth. This year, for the first time, we are elevating the employment-first policy to the status of a macro policy. This is to increase society-wide attention to employment and support for it. Both in the immediate future and for some time to come, the pressure on aggregate job creation will continue unabated, the related structural issues will become more pronounced, and new factors that affect employment will continue to grow. All this means we must give greater priority to increasing employment. Maintaining stable growth, first and foremost, is to ensure employment. This year, on top of the urban job creation target, we will work to reach the actual employment figures of the past few years so as to ensure employment for the urban workforce while creating non-agricultural employment opportunities for the surplus rural workforce. With stable employment and increasing incomes, we can continue to be fully confident.

注解
Notes

翻译过程中，有些地方需要作格式上的调整。比如文字可以翻译成表格样式，或表格样式也可以翻译成文字。在本单元中段落文字部分有列表样式翻译。

如汉语原文为：

今年是新中国成立70周年，是全面建成小康社会、实现第一个百年奋斗目标的关键之年。做好政府工作，要在以习近平同志为核心的党中央坚强领导下，以习近平新时代中国特色社会主义思想为指导，全面贯彻党的十九大和十九届二中、三中全会精神，统筹推进"五位一体"总体布局，协调推进"四个全面"战略布局，坚持稳中求进工作总基调，坚持新发展理念，坚持推动高质量发展，坚持以供给侧结构性改革为主线，坚持深化市场化改革、扩大高水平开放，加快建设现代化经济体系，继续打好三大攻坚战，着力激发微观主体活力，创新和完善宏观调控，统筹推进稳增长、促改革、调结构、惠民生、防风险、保稳定工作，保持经济运行在合理区间，进一步稳就业、稳金融、稳外贸、稳外资、稳投资、稳预期，提振市场信心，增强人民群众获得感、幸福感、安全感，保持经济持续健康发展和社会大局稳定，为全面建成小康社会收官打下决定性基础，以优异成绩庆祝中华人民共和国成立70周年。

英文译文处理如下：

This year is the 70th anniversary of the founding of the People's

Republic of China. It will be a crucial year for us as we endeavor to achieve the first Centenary Goal of building a moderately prosperous society in all respects. To fulfill the work of government, under the strong leadership of the Party Central Committee with Comrade Xi Jinping at its core, we must:

• follow the guidance of Xi Jinping Thought on Socialism with Chinese Characteristics for a New Era;

• implement fully the guiding principles of the Party's 19th National Congress and the second and third plenary sessions of its 19th Central Committee;

• pursue coordinated progress in the five-sphere integrated plan;

• pursue balanced progress in the four-pronged comprehensive strategy;

• adhere to the general principle of pursuing progress while ensuring stability;

• continue to apply the new development philosophy;

• continue to work for high-quality development;

• continue to pursue supply-side structural reform as our main task;

• continue to deepen market-orientated reforms and expand high-standard opening up;

• work faster to modernize the economy;

• continue the three critical battles;

• invigorate micro entities;

• explore innovations in and improve macro regulation;

• make coordinated efforts to maintain stable growth, advance reform, make structural adjustments, improve living standards, guard against risks and ensure stability.

• keep major economic indicators within an appropriate range;

• take further steps to ensure stable employment, a stable financial sector, stable foreign trade, stable foreign investment, stable domestic investment, and stable expectations;

• boost market confidence;

- enable people to feel more satisfied, happy, and secure;
- sustain healthy economic development and maintain social stability.

By doing the above, we will create the pivotal underpinning for completing the building of a moderately prosperous society and celebrate with outstanding accomplishments the 70th anniversary of the founding of the People's Republic of China.

练习
Exercises

练习一　将课文做英汉互译练习

练习二　阅读理解

Cyberspace, data superhighways, multi-media—for those who have seen the future, the linking of computers, televisions and telephones will change our lives forever. Yet for all the talk of a forthcoming technological utopia, little attention has been given to the implications of these developments for the poor. As with all new high technology, while the West concerns itself with the "how," the question of "for whom" is put aside once again.

Economists are only now realizing the full extent to which the communications revolution has affected the world economy, information technology allows the extension of trade across geographical and industrial boundaries, and transnational corporations take full advantage of it. Terms of trade, exchange and interest rates and money movements are more important than the production of goods. The electronic economy made possible by information technology allows the haves to increase their control on global markets—with destructive impact on the have-nots.

For them the result is instability. Developing countries which rely on the production of a small range of goods for export are made to feel like small parts in the international economic machine. As "futures" are traded on computer screens, developing countries simply have less and less control of their destinies.

So what are the options for regaining control? One alternative is for developing countries to buy in the latest computers and telecommunications themselves—so-called "development communications" modernization. Yet this leads to long-term dependency and perhaps permanent constraints on developing countries' economies.

Communications technology is generally exported from the U.S., Europe or Japan; the patents, skills and ability to manufacture remain in the hands of a few industrialized countries. It is also expensive, and imported products and services most therefore be bought on credit—credit usually provided by the very countries whose companies stand to gain.

Furthermore when new technology is introduced, there is often too low a level of expertise to exploit it for native development. This means that while local elites, foreign communities and subsidiaries of transnational corporations may benefit, those whose lives depend on access to the information are denied it.

1. From the passage we know that the development of high technology is in the interests of _____.
   A. the rich countries
   B. scientific development
   C. the elite
   D. the world economy

2. It can be inferred from the passage that _____.
   A. international trade should be expanded
   B. the interests of the poor countries have not been given enough consideration
   C. the exports of the poor countries should be increased
   D. communications technology in the developing countries should be modernized

3. Why does the author say that the electronic economy may have a destructive impact on developing countries?

A. Because it enables the developed countries to control the international market.

B. Because it destroys the economic balance of the poor countries.

C. Because it violates the national boundaries of the poor countries.

D. Because it inhibits the industrial growth of developing countries.

4. The development of modem communications technology in developing countries may _____.

A. hinder their industrial production

B. cause them to lose control of their trade

C. force them to reduce their share of exports

D. cost them their economic independence

5. The author's attitude toward the communications revolution is _____.

A. positive   B. critical

C. indifferent   D. tolerant

第四单元　中华人民共和国国民经济和社会发展第十三个五年规划纲要（节选）

UNIT FOUR
THE 13TH FIVE-YEAR PLAN FOR ECONOMIC AND SOCIAL DEVELOPMENT OF THE PEOPLE'S REPUBLIC OF CHINA (EXCERPTED)

第四篇　推进农业现代化
PART IV　AGRICULTURAL MODERNIZATION

农业是全面建成小康社会和实现现代化的基础，必须加快转变农业发展方式，着力构建现代农业产业体系、生产体系、经营体系，提高农业质量效益和竞争力，走产出高效、产品安全、资源节约、环境友好的农业现代化道路。

Agriculture is the foundation on which we can finish building a moderately prosperous society in all respects and achieve modernization. The agricultural growth model must be transformed at a faster pace, industrial, production and business operation systems that work for modern agriculture must be established, and the quality, returns, and competitiveness of agriculture must be strengthened to allow China to embark on a path of modern agricultural development which ensures high yields and safe products, conserves resources, and is environmentally friendly.

第十八章　增强农产品安全保障能力
Chapter 18　Strengthen Capacity for Ensuring Safety of Agricultural Products

确保谷物基本自给、口粮绝对安全，调整优化农业结构，提高农产品综合生产能力和质量安全水平，形成结构更加合理、保障更加有力的农产品有效供给。

We will make sure we achieve basic self-sufficiency in cereal grains and absolute food security, make agricultural structural adjustments and improvements, raise our production capacity for agricultural products while also improving quality and safety, and see that a better structured, more effective supply of agricultural products takes shape.

第一节　提高粮食生产能力保障水平
Section 1　Safeguards for Grain Production

坚持最严格的耕地保护制度，全面划定永久基本农田。实施藏粮于地、藏粮于技战略，以粮食等大宗农产品主产区为重点，大规模推进农田水利、土地整治、中低产田改造和高标准农田建设。完善耕地占补平衡制度，研究探索重大建设项目国家统筹补充耕地办法，全面推进建设占用耕地耕作层剥离再利用。建立粮食生产功能区和重要农产品生产保护区，确保稻谷、小麦等口粮种植面积基本稳定。健全粮食主产区利益补偿机制。深入推进粮食绿色高产高效创建。

We will continue to apply the strictest possible protection system for farmland and will designate permanent basic cropland throughout the country. We will put in place a food crop production strategy that is based on farmland management and the application of technology, and with the focus on major growing areas of grain crop and other staple agricultural products, we will make a large-scale push to see the building of farmland irrigation systems and water conservancy infrastructure, the restoration of rural land, the improvement of low-and medium-yield cropland, and the development of high-quality farmland. We will improve the system for ensuring that cultivated land taken over for nonagricultural use is replaced with land of an equivalent amount and quality, explore the possibility of formulating measures for national coordination in offsetting farmland that has been used for major construction projects, and ensure that the stripping and reuse of topsoil are practiced nationwide on all cultivated land put to nonagricultural use. We will establish grain crop production functional zones and protected areas for the production of major agricultural products to ensure that the arrange of land devoted to growing grain crops such as rice and wheat remains basically stable. We will improve the mechanism for subsidizing major grain crop production areas. We will intensify efforts to realize green, high yield, and efficient grain crop production.

第二节 加快推进农业结构调整
Section 2 Agricultural Structural Adjustments

推动粮经饲统筹、农林牧渔结合、种养加一体发展。积极引导调整农业种植结构，支持优势产区加强棉花、油料、糖料、大豆、林果等生产基地建设。统筹考虑种养规模和资源环境承载力，推广粮改饲和种养结合模式，发展农区畜牧业。分区域推进现代草业和草食畜牧业发展。提高畜禽、水产标准化规模化养殖水平。促进奶业优质安全发展。实施园艺产品提质增效工程。发展特色经济林和林下经济。优化特色农产品生产布局。加快现代农业示范区建设。

We will promote the coordination of food, cash, and fodder crop production, step up integrated development of the farming, forestry, livestock and fishing industries, and integrate planting, breeding and processing. We will actively guide adjustments to the production mix of agricultural products and support superior producing areas in developing production centers for cotton, oilseed, sugar crops, soybeans, forestry seedlings and fruit. Taking into consideration the scale of planting and breeding operations and resource and environmental carrying capacities, we will promote models that swap food crop for fodder crop cultivation and integrate planting and breeding operations, and develop the farming area-based livestock industry. We will develop modern grass industries and herbivorous livestock industries region by region. We will ensure that livestock, poultry, and aquaculture farming are further standardized and brought up to scale. We will ensure that the dairy industry produces safe, quality products. We will improve the quality and cost-effectiveness of horticultural products. We will develop economic forests and under-forest economies that take advantage of local strengths. We will improve the geographical layout of the production of specialty agricultural products. We will accelerate the development of demonstration areas for modern agriculture.

第三节 推进农村一二三产业融合发展
Section 3　Integrated Development of Primary, Secondary and Tertiary Industries in Rural Areas

推进农业产业链和价值链建设，建立多形式利益联结机制，培育融合主体、创新融合方式，拓宽农民增收渠道，更多分享增值收益。积极发展农产品加工业和农业生产性服务业。拓展农业多种功能，推进农业与旅游休闲、教育文化、健康养生等深度融合，发展观光农业、体验农业、创意农业等新业态。加快发展都市现代农业。激活农村要素资源，增加农民财产性收入。

We will promote the development of agricultural production and value chains, create different types of linkages between the interests of different entities, and foster entities that integrate primary, secondary and tertiary industry operations as well as help develop new kinds of such operations in order to open up more channels through which rural residents can increase their incomes and enable them to benefit more from the resultant value-added. We will promote the development of processing industries and services for agricultural production. We will see that agriculture takes on more functions, promote close cooperation between agriculture on the one hand and leisure, tourism, education, culture, and health on the other, and develop new forms of agricultural business such as agri-tourism, agricultural experiences, and creative agriculture. We will accelerate the development of modern urban agriculture. We will put rural resources and factors of production to better use so as to increase the property income of rural residents.

第四节 确保农产品质量安全
Section 4　Agricultural Product Quality and Safety

加快完善农业标准，全面推行农业标准化生产。加强农产品质量安全和农业投入品监管，强化产地安全管理，实行产地准出和市场准入制度，建立全程可追溯、互联共享的农产品质量安全信息平台，健全从农田到餐桌的农产品质量安全全过程监管体系。强化农药和兽药残留超标治理。严格食用农产品添加剂控制标准。开展国家农产品质量安全县创建行动。加强动植物疫病防控能力建设，强化进口农产品质量安全监管。创建优质农产品品牌，支持品牌化营销。

We will move faster to improve agricultural standards and ensure they are met in all agricultural production. We will strengthen quality and safety oversight over agricultural products and inputs, strengthen safety management at production sites, implement a certification system for product oversight both before products leave production sites and on entrance into markets, and establish interconnected, shared agricultural product quality and safety information platforms allowing for full traceability at every stage, thereby forming a stronger quality and safety oversight system that covers the entire journey of agricultural products from farm to table. We will strengthen work on addressing excess residues of pesticides and livestock medicines. We will enforce strict standards for the control of additives in edible agricultural products. We will move forward with the national initiative to develop counties with advanced capabilities for ensuring agricultural product quality and safety. We will strengthen our ability to prevent and control animal and plant diseases and increase oversight over the quality and safety of agricultural imports. We will develop brands of quality agricultural products and support brand marketing.

第五节　促进农业可持续发展
Section 5　Sustainable Agricultural Development

大力发展生态友好型农业。实施化肥农药使用量零增长行动，全面推广测土配方施肥、农药精准高效施用。实施种养结合循环农业示范工程，推动种养业废弃物资源化利用、无害化处理。开展农业面源污染综合防治。开展耕地质量保护与提升行动，推进农产品主产区深耕深松整地，加强东北黑土地保护。重点在地下水漏斗区、重金属污染区、生态严重退化地区，探索实行耕地轮作休耕制度试点。在重点灌区全面开展规模化高效节水灌溉行动。推广旱作农业。在南疆叶尔羌河、和田河等流域，以及甘肃河西走廊、吉林白城等严重缺水区域，实施专项节水行动计划。加强气象为农服务体系建设。创建农业可持续发展试验示范区。

We will work hard to develop eco-friendly agriculture. We will carry out the initiative to achieve zero growth in the use of chemical fertilizer and pesticides and promote fertilizer use based on the results of soil tests as well as the targeted and effective use of pesticides nationwide. We will implement a demonstration project for circular agriculture through integrated planting and breeding, and promote the recovery of resource and safe disposal of waste materials from planting and breeding industries. We will take comprehensive measures to prevent and control agricultural pollution from non-point sources. We will protect and improve the quality of cultivated land, promote deep planting and plowing to improve cropland in major agricultural product production areas, and strengthen the protection of chernozem soil in northeast China. We will pilot crop rotation and fallow systems focusing on cones of depression, heavy metal contaminated areas and areas suffering serious ecological degradation. Large-scale, high efficiency, and water-saving irrigation will be introduced in all key irrigation areas. Dry farming will be encouraged in more areas, Special action plans for promoting water conservation will be implemented in the drainage basins of Yarkand, Hotan, and other rivers in southern Xinjiang and in seriously water-deprived areas such as the Hexi Corridor of Gansu and Baicheng in Jilin. We will strengthen development of the meteorological service system for agriculture. We will establish pilot demonstration zones for sustainable agricultural development.

第六节　开展农业国际合作
Section 6　International Cooperation in Agriculture

健全农产品贸易调控机制，优化进口来源地布局，在确保供给安全条件下，扩大优势农产品出口，适度增加国内紧缺农产品进口。积极开展境外农业合作开发，建立规模化海外生产加工储运基地，培育有国际竞争力的农业跨国公司。拓展农业国际合作领域，支持开展多双边农业技术合作。

We will improve mechanisms for regulating trade in agricultural products, optimize the mix of sources of imports, expand exports of competitive agricultural products while ensuring domestic supply, and appropriately increase imports of agricultural products that are in short supply at home. We will actively pursue agricultural cooperation and development overseas, establish large-scale offshore centers for farm product production, processing, storage, and transportation, and cultivate internationally competitive multinational agricultural companies. We will broaden the areas of international agricultural cooperation and support bilateral and multilateral cooperation in agricultural technology.

第十九章　构建现代农业经营体系
Chapter 19　Establish a Modern Agricultural Operations System

以发展多种形式适度规模经营为引领，创新农业经营组织方式，构建以农户家庭经营为基础、合作与联合为纽带、社会化服务为支撑的现代农业经营体系，提高农业综合效益。

Guided by the need to develop different forms of appropriately scaled agricultural operations, we will create new methods for organizing agricultural operations and establish a modern system of agricultural operations that is based on rural household operations, held together by cooperation and association, and supported by society-wide services thereby increasing the overall returns of agriculture.

第一节　发展适度规模经营
Section 1　Appropriately Scaled Agricultural Operations

稳定农村土地承包关系，完善土地所有权、承包权、经营权分置办法，依法推进土地经营权有序流转，通过代耕代种、联耕联种、土地托管、股份合作等方式，推动实现多种形式的农业适度规模经营。

We will keep rural land contract relationships stable, improve the measures for separating land ownership rights, contract rights, and management rights, promote the orderly transfer of land management rights in accordance with the law, and develop appropriately scaled agricultural operations through means such as third party cultivation, joint cultivation of combined land, land trusteeship, and joint-stock cooperation

第二节　培育新型农业经营主体
Section 2　New Types of Agribusiness

健全有利于新型农业经营主体成长的政策体系，扶持发展种养大户和家庭农场，引导和促进农民合作社规范发展，培育壮大农业产业化龙头企业，大力培养新型职业农民，打造高素质现代农业生产经营者队伍。鼓励和支持工商资本投资现代农业，促进农商联盟等新型经营模式发展。

We will make policies more conducive to the growth of new types agribusiness, support the development of large family farming businesses and family farms, guide and promote the well-regulated development of farmers' cooperatives, support the growth of enterprises that are leaders in agricultural industrialization, cultivate a new type of professional farmer, and nurture competent modern agricultural operators. We will encourage and support industrial and commercial capital investment in modern agriculture and promote the development of agricultural-commercial alliances and other emerging business models.

第三节　健全农业社会化服务体系
Section 3　The System of Society-Wide Services for Agriculture

实施农业社会化服务支撑工程，培育壮大经营性服务组织。支持科研机构、行业协会、龙头企业和具有资质的经营性服务组织从事农业公益性服务，支持多种类型的新型农业服务主体开展专业化、规模化服务。推进农业生产全程社会化服务创新试点，积极推广合作式、托管式、订单式等服务形式。加强农产品流通设施和市场建设，完善农村配送和综合服务网络，鼓励发展农村电商，实施特色农产品产区预冷工程和"快递下乡"工程。深化供销合作社综合改革。创新农业社会化服务机制。

We will implement a program to support society-wide services for agriculture, and develop commercial service organizations. We will support the provision of public-benefit agricultural services by research institutions, industry associations, leading enterprises, and qualified commercial service organizations, and support different types of new agricultural service entities in offering professional services on a large scale. We will give impetus to trials for making innovations in society-wide services at every stage of the agricultural production process, and actively promote various forms of services such as cooperative, trusteeship-based, and order-based services. We will strengthen the development of distribution facilities and markets for agricultural products, work to improve rural logistics and comprehensive service networks, encourage the development of e-commerce in rural areas, and implement projects to encourage the development of precooling in specialty agricultural product producing areas and express delivery services in rural areas. We will deepen the comprehensive reform of supply and marketing cooperatives. We will make innovations in society-wide agricultural service mechanisms.

第二十章　提高农业技术装备和信息化水平
Chapter 20　Improve Technology and Equipment and Increase Information Technology Application in Agriculture

健全现代农业科技创新推广体系，加快推进农业机械化，加强农业与信息技术融合，发展智慧农业，提高农业生产力水平。

With the aim of raising agricultural productivity, we will improve system for promoting innovation in and the application of modern agricultural science and technology, accelerate agricultural modernization, strengthen the integration of information technology into agriculture, and develop intelligent agriculture.

第一节 提升农业技术装备水平
Section 1 Agricultural Technology and Equipment

加强农业科技自主创新,加快生物育种、农机装备、绿色增产等技术攻关,推广高产优质适宜机械化品种和区域性标准化高产高效栽培模式,改善农业重点实验室创新条件。发展现代种业,开展良种重大科技攻关,实施新一轮品种更新换代行动计划,建设国家级育制种基地,培育壮大育繁推一体化的种业龙头企业。推进主要作物生产全程机械化,促进农机农艺融合。健全和激活基层农业技术推广网络。

We will strengthen innovation in agricultural science and technology and accelerate work on developing bio-breeding, agricultural machinery and equipment, and eco-friendly methods for increasing production. We will promote the use of high-yield, high-quality crop breeds suited to mechanized agriculture as well as standardized and localized models of high-yield and high-performance cultivation, and we will improve the conditions for making innovations in major agricultural laboratories. We will develop the modern seed industry, tackle key scientific and technological issues to make progress in the development of superior seed varieties, implement a new action plan for upgrading crop varieties, develop national seed breeding and production centers, and help the growth of leading seed enterprises using integrated cultivation-breeding-promotion operations. We will promote complete mechanization of the production process of major crops as well as the integration of agricultural machinery and methods. We will improve and invigorate networks for the promotion of agricultural technology at the community level.

第二节 推进农业信息化建设
Section 2 Information Technology Adoption in Agriculture

推动信息技术与农业生产管理、经营管理、市场流通、资源环境等融合。实施农业物联网区域试验工程,推进农业物联网应用,提高农业智能化和精准化水平。推进农业大数据应用,增强农业综合信息服务能力。鼓励互联网企业建立产销衔接的农业服务平台，加快发展涉农电子商务。

We will promote the integration of information technology into agricultural production management, operations management, market distribution, and fields related to resources and the environment. We will help spread the Internet of Things into agriculture by carrying out an experimental project to promote its use in certain regions, thereby promoting the development of intelligent agriculture and precision agriculture. We will promote the use of big data in agriculture and strengthen the overall capabilities of agricultural information services. We will encourage internet enterprises to establish agricultural service platforms that bring together the processes of production and marketing, and accelerate the development of agriculture-related e-commerce.

第二十一章　完善农业支持保护制度
Chapter 21　Improve Systems for Providing Support and Protection for Agriculture

以保障主要农产品供给、促进农民增收、实现农业可持续发展为重点，完善强农惠农富农政策，提高农业支持保护效能。

With an emphasis on ensuring the supply of major agricultural products, promoting increases in rural incomes, and achieving sustainable agricultural development, we will improve policy support aimed at strengthening agriculture, benefiting farmers, and raising rural living standards and raise our level of support and protection for agriculture.

第一节　持续增加农业投入
Section 1　Increased Investment in Agriculture

建立农业农村投入稳定增长机制。优化财政支农支出结构，创新涉农资金投入方式和运行机制，推进整合统筹，提高农业补贴政策效能。逐步扩大"绿箱"补贴规模和范围，调整改进"黄箱"政策。将农业"三项补贴"合并为农业支持保护补贴，完善农机具购置补贴政策，向种粮农民、新型经营主体、主产区倾斜。建立耕地保护补偿制度。

We will establish a mechanism for steadily increasing investment in agriculture and rural areas. In the area of agricultural investment, we will improve the government spending mix, create new ways of investing and operating government funds, promote the integration of investment projects, and improve the efficacy of subsidy policies. We will progressively increase the range and scale of green box subsidies while adjusting and improving amber box policies. The subsidies for food crop production, for promoting superior grain crop varieties, and for supporting the purchase of agricultural supplies will be combined into a single agricultural support and protection subsidy. We will improve subsidy policies for the purchase of agricultural machinery and tools, and give priority to grain crop producers, new types of agribusinesses, and major agricultural production areas in the allocation of these subsidies. We will establish a system of protection and compensation for arable land.

第二节 完善农产品价格和收储制度
Section 2　Pricing, Purchasing, and Stockpiling Systems for Agricultural Products

坚持市场化改革取向和保护农民利益并重，完善农产品市场调控制度和市场体系。继续实施并完善稻谷、小麦最低收购价政策。深化棉花、大豆目标价格改革。探索开展农产品目标价格保险试点。积极稳妥推进玉米价格形成机制和收储制度改革，建立玉米生产者补贴制度。实施粮食收储供应安全保障工程，科学确定粮食等重要农产品储备规模，改革完善粮食储备管理体制和吞吐调节机制，引导流通、加工企业等多元化市场主体参与农产品收储。推进智慧粮库建设和节粮减损。

We will ensure equal emphasis is placed on both carrying out market-oriented reforms and protecting the interests of farmers, and improve the system for regulating the market for agricultural products and the market system itself. We will continue to implement and improve the minimum purchase price policy for rice and wheat, and deepen reform of the program for guaranteeing base prices for cotton and soybeans. We will explore the possibility of trialing base price insurance for agricultural products. We will actively and prudently carry out reform of the price-setting mechanism and the purchasing and stockpiling systems for corn, and establish a system for subsidizing corn producers. We will implement a project to ensure security of the purchase, stockpiling, and supply of grain crops, research and determine the optimum scale of reserves for grain crops and other important agricultural products, reform and improve the grain crop reserve management system as well as mechanisms for grain crop regulation and adjustment, and guide a diverse range of market entities-such as distribution and processing businesses-in participating in the purchase and stockpiling of agricultural products. We will move forward with the development of intelligent storage facilities for grain crops and work to conserve grain crops and reduce waste.

第三节 创新农村金融服务
Section 3 Innovations in Rural Financial Services

发挥各类金融机构支农作用，发展农村普惠金融。完善开发性金融、政策性金融支持农业发展和农村基础设施建设的制度。推进农村信用社改革，增强省级联社服务功能。积极发展村镇银行等多种形式农村金融机构。稳妥开展农民合作社内部资金互助试点。建立健全农业政策性信贷担保体系。完善农业保险制度，稳步扩大"保险＋期货"试点，扩大保险覆盖面，提高保障水平，完善农业保险大灾风险分散机制。

We will ensure all types of financial institutions support agriculture, and develop inclusive financing in rural areas. We will improve systems for supporting agricultural development and for the construction of rural infrastructure using development and policy-backed financing We will carry out reform of rural credit cooperatives and strengthen the service functions of provincial-level unions of such cooperatives. We will actively develop diverse forms of rural financial institutions such as village banks. We will steadily carry out trials to allow farmers' cooperatives to use internal funds to provide financial services to their members. We will establish a sound policy-backed credit guaranty system for agriculture. We will improve the agricultural insurance system, steadily increase trials of the "insurance + futures" model, expand the scope of insurance coverage, raise insurance benefits and improve the risk-spreading mechanisms of agricultural insurance against the risks posed by major disasters.

注解
Notes

汉语原文许多句子为无主句,英译时添加适当主语。

汉语原文为:

坚持最严格的耕地保护制度,全面划定永久基本农田。实施藏粮于地、藏粮于技战略,以粮食等大宗农产品主产区为重点,大规模推进农田水利、土地整治、中低产田改造和高标准农田建设。完善耕地占补平衡制度,研究探索重大建设项目国家统筹补充耕地办法,全面推进建设占用耕地耕作层剥离再利用。建立粮食生产功能区和重要农产品生产保护区,确保稻谷、小麦等口粮种植面积基本稳定。健全粮食主产区利益补偿机制。深入推进粮食绿色高产高效创建。

以上段落,"坚持……制度""实施……战略""完善……制度""研究探索""建立""健全"等都是无主句句式,英文适当添加主语"We"。以下为该段落中文的英语译文。

We will continue to apply the strictest possible protection system for farmland and will designate permanent basic cropland throughout the country. We will put in place a food crop production strategy that is based on farmland management and the application of technology, and with the focus on major growing areas of grain crop and other staple agricultural products, we will make a large-scale push to see the building of farmland irrigation systems and water conservancy infrastructure, the restoration of rural land, the improvement of low-and medium-yield cropland, and the development of high-quality farmland. We will improve the system for ensuring that cultivated land taken over for nonagricultural use is replaced with land of an equivalent amount and quality, explore the possibility of formulating measures for national coordination in offsetting farmland that has been used for major construction projects, and ensure that the stripping and reuse of topsoil are practiced nationwide on all cultivated land put to nonagricultural use. We will establish grain crop production functional zones and protected areas for the production of major agricultural products to ensure that the arrange of land devoted to growing grain crops such as rice and wheat remains basically stable. We will improve the mechanism for subsidizing major grain crop production areas. We will intensify efforts to realize green, high yield, and efficient grain crop production.

练习
Exercises

练习一　将课文内容做英汉互译练习

练习二　阅读理解

Not content with its doubtful claim to produce cheap food for our own population, the factory farming industry also argues that "hungry nations are benefiting from advances made by the poultry industry". In fact, rather than helping the fight against malnutrition in "hungry nations", the spread of factory farming has, inevitably, aggravated the problem.

Large-scale intensive meat and poultry production is a waste of food resources. This is because more protein has to be fed to animals in the form of vegetable matter than can ever be recovered in the form of meat. Much of the food value is lost in the animal's process of digestion and cell replacement. Neither, in the ease of chicken, can one eat feathers, blood, feet or head. In all, only about 44% of the live animal fits to be eaten as meat.

This means one has to feed approximately 9—10 times as much food value to the animal as one can consume from the carcass. As a system for feeding the hungry, the effects can prove disastrous. At times of crisis, grain is the food of life.

Nevertheless, the huge increase in poultry production throughout Asia and Africa continues. Normally British or US firms are involved. For instance, an American based multinational company has this year announced its involvement in projects in several African countries. Britain's largest suppliers chickens, Ross Breeders, are also involved in projects all over the world.

Because such trade is good for exports, Western governments encourage it. In 1979, a firm in Bangladesh called Phoenix Poultry received a grant to set up a unit of 6,000 chickens and 18,000 laying hens. This almost doubled the number of poultry kept in the country all at once.

But Bangladesh lacks capital, energy and food and has large numbers of unemployed. Such chicken-raising demands capital for building and machinery, extensive use of energy resources for automation, and involves feeding chickens with potential famine-relief protein food. At present, one of Bangladesh's main imports is food grains, because the country is unable to grow enough food to feed its population. On what then can possible feed the chicken?

1. In this passage the author argues that _____.
A. efficiency must be raised in the poultry industry
B. raising poultry can provide more protein than growing grain
C. factory farming will do more harm than good to developing countries
D. hungry nations may benefit from the development of the poultry industry

2. According to the author, in factory, vegetable food _____.
A. is easy for chickens to digest
B. is insufficient for the needs of poultry
C. is fully utilized in meat and egg production
D. is inefficiently converted into meat and eggs

3. Western governments encourage the poultry industry in Asia because they regard it as an effective way to _____.
A. boost their own exports
B. alleviate malnutrition in Asian countries

C. create job opportunities in Asian countries

D. promote the exports of Asian countries

4. The word "carcass" (Line 2, Para 3) most probably means _____.

A. vegetables preserved for future use

B. the dead body of an animal ready to be cut into meat

C. expensive food that consumers can hardly afford

D. meat canned for future consumption

5. What the last paragraph tells us is the author's _____.

A. detailed analysis of the ways raising poultry in Bangladesh

B. great appreciation of the development of poultry industry in Bangladesh

C. critical view on the development of the poultry industry in Bangladesh

D. practical suggestion for the improvement of the poultry industry in Bangladesh

第二部分 农林科普

PART TWO POPULAR SCIENCES OF AGRICULTURE AND FORESTRY

第二部分 农林科普

第一单元 研发超级稻
UNIT ONE
THE SUPER RICE CHALLENGE

Rice is the main food for about one-third to one-half of the world's population. A mature rice plant is usually two to six feet tall. In the beginning, one shoot appears. It is followed by one, two, or more offshoots developing. There are at least five or six hollow joints for each stalk, and a leaf for each joint. The leaf of the rice plant is long, flat, and stiff. The highest join of the rice plant is called the panicle. The rice grains develop from the panicles. Rice is classified in the grass family Gramineae. Its genus is Oryza and species O. sativa. It is commonly cultivated for food in Asia. Some varieties of rice include red rice, glutinous rice, and wild rice.

The kernel within the grain contains most of the vitamins and minerals. The kernel contains thiamine, niacin, and riboflavin. Rice has many enemies that destroy a majority of the rice crops. The larvae of moth, stem borers, live in the stems of the rice plants. Some insects suck the plant juices or chew the leaves. Birds, such as bobolink, Java sparrow, or paddy bird, would eat the seeds or grains. Disease causing factors such as fungi, roundworms, viruses, and bacteria also destroy the rice plants. Blast disease is caused by fungi which cause the panicles containing the grains to break.

The purpose of the Super Rice challenge is to create rice plants that are disease resistant, insect resistant, and produces twenty-five percent more food per acre. The International Rice Research Institute has been working on this challenge. It is competing with many various factors that are pushing the International Rice Research Institute to try and complete the challenge as soon as possible. Factors such as growing population, limited areas for growing rice, and the common farmer's philosophy of getting anything to grow are pushing researchers to complete the challenge as soon as possible. Also, the new varieties of rice have raised a question of the farmer's health because of the uses and effects of agricultural chemicals. Since normal rice grown in paddies produces high amounts of methane, the International Rice Research Institute must also find a way to create rice plants with a low methane production.

Gurdev Khush believes that the super rice will be ready for many farmers to plant them early this century. Researchers were able to develop a type of rice during the 1960's. This type of rice is called miracle rice because of its high yields. Researchers were able to develop it by combining a short variety of rice with a tall variety. This crossbreeding resulted in a rice plant that can withstand wind and rain and have a high production yield. This new breed was thought to have reduced the food shortages that depend on rice as a staple food but because of various conditions in other countries, this rice plant was not very successful. Blight, caused by bacteria, spread rapidly through rice fields in water droplets. The rice plant would develop lesions and die in a matter of days. This disease could destroy about half of a rice crop. Through genetic engineering, the author and her colleagues have been able to introduce isolated disease-resistant genes into the rice plants. The gene, called $xa21$, was discovered by the International Rice Research Institute, and Ronald attempted to clone $xa21$ from the International Rice Research Institute variety. The Cornell group created a genetic map which showed the location of hundreds of markers on the twelve rice chromosomes. Ronald and her colleagues used this genetic map to locate the gene $xa21$ by examining over one thousand rice plants to see how often known DNA markers showed up in conjunction with resistance to blight. They used chromosomal swapping and rearranging that goes on during sexual reproduction. The more often they saw resistance in the next generation of rice plants, the closer they were to locating the gene. Since rice plants are defiant in accepting outside DNA, they used a gun that shoots microscopic particles into intact cells which was developed by John Sanford of Cornell. After using this procedure to introduce $xa21$ into an old, but susceptible, rice plant they exposed the plants to blight. They found that the plants were resistant to the blight. The current goal of Ronald and her colleagues is to introduce $xa21$ into rice varieties that are agriculturally important.

Current studies showed that rice plants introduced to the cloned $xa21$ gene have become blight resistant. Since farmers prefer to grow plants that have adapted to the various climates and conditions, Ronald stated that the engineered versions will be identical to the original plants except for the addition of the single cloned gene. Ronald and her colleagues still have to field-test the new varieties for yield, taste, and hardiness to confirm that the original adaptations have remained unchanged. The success of this project has reached into testing the process and the gene on other plants. Scientists hope that Ronald's process of making the rice plant blight resistant will work on other plants. They hope that this process will be successful on valuable crops, such as citrus crops. They plan to combine the gene $xa21$ and other disease resistance genes to enhance the plant's resistance to disease. The problem with cloning the $xa21$ gene is that it is still vulnerable to other diseases such as grassy-stunt and ragged-stunt viruses.

The purpose of Japan's rice genome project is to fully map the twelve chromosomes of the rice plant. Low funding of this project has hindered the progression of this project. Since Japan has increased its funding to its genome project, the rice genome division can now complete mapping the twelve chromosomes of the rice plant. Rice is one of the world's most important crops because a majority of the world depends on this as a staple food. The number of rice plants planted, however, is greater than the number of rice consumed. This is because of various factors that destroy the rice plants before they can be harvested for commercial use. Various factors, such as insects, birds, and disease, destroy the rice crops. Projects are being conducted to improve the rice plant, but researchers encounter various obstacles. Making the rice plant disease-resistant to blight may be useful and valuable, but they must also find a way to make the rice plant resistant to other diseases and viruses such as ragged-stunt. Since Japan has increased its funds to its genome projects, they have been able to increase work on mapping the twelve rice chromosomes. Scientists hope that these projects will be finished, and that farmers will be using the enhanced genes on their rice very soon.

生词和词组
New Words and Expressions

stalk	（植）茎，秆子
panicle	（植）圆锥状花序
Gramineae	禾本科
kernel	谷物或坚果的仁
thiamine	维生素 B_1，硫胺
niacin	烟碱酸，抗癞皮病维生素
riboflavin	核黄素；维生素 B_2
bobolink	食米鸟
fungi	（fungus 的复数）真菌
paddy	水稻田
methane	沼气，甲烷
conjunction	连接，同时发生
blight	（植）枯死病；招致毁灭或破坏的原因
swap	交换
citrus	柑橘类植物

专有名词
Proper nouns

The International Rice Research Institute	国际水稻研究所
Gurdev Khush	（人名）戈德夫·库什
Ronald	（人名）罗纳德
The Cornell group	康奈尔大学实验组
John Stanford	（人名）约翰·司丹佛

注解
Notes

1. Ronald and her colleagues used this genetic map to locate gene $xa21$ by examining over one thousand rice plants to see how often known DNA markers showed up in conjunction with resistance to blight.

这是一个主从复合句，主句中第一个不定式短语 to locate... 作宾语补足语；第二个不定式短语 to see... 作目的状语，该不定式还带一个 how 引导的宾语从句。

2. Since rice plants are defiant in accepting outside DNA, they used a gun that shoots microscopic particles into intact cells, which was developed by John Sanford of Cornell.

在这个主从复合句中，Since 引导的是一个原因状语从句，which 引导的是一个非限定性定语从句，修饰整个主句。

3. grassy-stunt and ragged-stunt viruses 为"水稻草矮病毒和水稻齿矮病毒"。

译文参考
Translation for Reference

研发超级稻

水稻是世界上 1/3 至 1/2 人口的主食。水稻的成株一般有 0.6 米～1.8 米高。起初，先是有一只芽冒了出来，接着又有一只、两只……更多的芽长了出来。每根稻秆上至少有五六个中空的节，每个节上长着一片叶子。水稻植株的叶子长而尖，平而硬，最高处的连接点叫作圆锥花序，稻粒就是从圆锥花序上长出来的。水稻属于草棵植物中的禾本科，在亚洲作为粮食作物被广泛种植。水稻的品种包括红米、糯米和野生稻米。

稻谷中包着的米仁含有大部分维生素和矿物质，米仁中含有维生素 B_1、烟碱酸、维生素 B_2。水稻有很多能够将其大部分作物毁坏掉的敌人。蛾子的幼虫、水稻螟虫，生长在水稻植株的茎里；一些昆虫吸食水稻的汁液或者吃水稻的叶子，像食米鸟、爪哇雀或者稻田鸟这样的鸟类会吃掉种子或稻谷；而像真菌、线虫、病毒、细菌这样的致病因素也会毁掉水稻植株，枯死病就是由导致包含稻谷的圆锥花序碎裂的真菌引发的。

超级稻研发的目的是为了创造出能抵御病虫害、每英亩增产25%以上的水稻。国际水稻研究所一直在进行这个挑战性的工作，有许多的各种各样的因素正在推动国际水稻研究所努力地尽早完成这个挑战。日益增长的人口、有限的水稻耕地以及像普通农民那样有东西可种的观点，这些因素都在推动研究者们及早完成这个项目。同时新品种水稻也提出了关于农药和化肥的使用和影响对农民健康造成的问题。因为正常种植在稻田里的水稻会产生大量的沼气，国际水稻研究所还得找出方法创造出一种沼气发生量低的水稻。

戈德夫·库什相信，21世纪初很多农民们可望种上超级稻。研究者们在20世纪60年代就已能开发出一种因其高产而被称为奇迹稻的水稻。这种水稻是研究人员用矮秆稻和高秆稻杂交培育出来的，这项杂交产生了一种能够抵挡风雨的高产水稻。人们本以为这个新品种能减少以大米为主食地区的食物短缺，但由于其他国家的不同情况，该品种并不太成功。细菌引起的枯死病由水滴携带着在整个稻田里迅速蔓延开来，水稻植株受到感染之后不过几天就会死掉，这种疾病将摧毁大约一半的水稻作物。通过基因工程，作者与其同事已能够将分离出来的抗病基因引入水稻植株中。这种叫作 $xa21$ 的基因是国际水稻研究所发现的，罗纳德试图从国际水稻研究所的品种中克隆 $xa21$。康奈尔大学实验组造出了一幅基因图，上面显示出水稻12条染色体上数以百计的基因标记的位置。罗纳德和她的同事们用这张图来寻找 $xa21$ 基因，他们仔细检查了1000多株水稻来弄清楚已知的DNA标记与抗枯死病关联出现的频率。他们使用了有性繁殖中的染色体交换和重新排序技术，下一代水稻植株中抗病基因出现的次数越高，离他们要找的基因就越近。因为水稻不接纳外来DNA，他们就使用康奈尔大学的约翰·司丹佛发明的方法，用一种枪将显微镜下才可见的粒子射入完整的细胞中。用这些步骤将 $xa21$ 引入一种既有的不具备免疫能力的水稻植株中之后，他们让这些植株接触有枯死病菌的植株，结果发现这些植株能够抵抗枯死病。罗纳德和她的同事们现在的目标是将 $xa21$ 引入那些具有重要农业价值的水稻品种中去。

目前的研究表明，那些引入了克隆的 $xa21$ 基因的水稻已经能够抵抗枯死病了。既然农民们更愿意种植适应本地气候及环境的品种，罗纳德说，经过基因改造的品种除了引入了这个克隆而来的基因外，其他均与其原品种毫无二致。罗纳德和她的同事们还得进行田间试验，以确定这些改良后的品种在产量、口味、抵抗力方面与其原品种的确一致。这个项目的成功已经延伸到了测试其他植物的方法和基因上，科学家们希望罗纳德的使水稻获得抗枯死病能力的方法能够在其他植物上奏效，他们希望这个方法在一些有价值的作物上，比如柑橘类作物，能够取得成功。他们打算将 $xa21$ 基因和其他抗病基因结合起来以提高植物的抗病能力。克隆 $xa21$ 基因存在着一个问题，就是在其他疾病，比如水稻草矮病毒和水稻齿矮病毒面前，它还很脆弱。

日本水稻基因组项目的目的是要完整绘出水稻的 12 条染色体的图，资金短缺曾使这个项目滞步不前。不过既然日本已经增加了对其基因组项目的资金投入，水稻基因组部门现在就能够完成对 12 条染色体的绘图了。水稻是世界上最重要的作物之一，因为世界上大多数人口以其为主食。但是，水稻的种植量是超过大米的消耗量的，这是因为在水稻能被收割下来出售之前遭受了各种各样的因素的破坏，这些破坏因素包括害虫、鸟以及疾病。正在开展的各种研究项目都是为了改进水稻，但是研究者们却遇到了各种各样的困难。使水稻植株具有抗枯死病的能力可能是有用的，也是有价值的，但是他们也必须找到办法来让水稻具备抵抗其他的像水稻齿矮这样的疾病和病毒的能力。日本已经增加了对其基因组项目的资金投入，他们已经能够在绘制水稻的 12 条染色体基因图上开展更多的工作。科学家们希望很快完成这些项目，让农民们种上基因改进过的水稻。

练习
Exercises

练习一　将课文内容做翻译练习

练习二　阅读理解

Passage A

Observe the dilemma of the fungus（真菌）: it is a plant, but it possesses no chlorophyll（叶绿素）while all other plants put the sun's energy to work for them combining the nutrients of ground and air into body structure, the chlorophylless fungus must look elsewhere for an energy supply. It finds it in those other plants which, having received theirs free from the sun, relinquish it at some point in their circle either to other animals or to fungi. In this search for energy the fungus has become the earth's major source of rot and decay. Wherever you see mold forming on a piece of bread, or a pile of leaves turning to compost（堆肥）, or a blown-down tree becoming pulp on the ground, you are watching a fungus eating. Without fungus action the earth would be piled high with the dead plant life of past centuries. In fact certain plants which contain resins（树脂，松香） that are toxic to fungi will last indefinitely; specimens of the redwood, for instance can still be found resting on the forest floor centuries after having been blown down.

1. The title below that best expresses the ideas of this passage is _____.

    A. Life without Chlorophyll

    B. The Strange World of the Fungus

    C. The Harmful Qualities of Fungi

    D. Utilization of the Sun's Energy

2. The statement "you are watching a fungus eating" is best described as _____.

    A. figurative   B. ironical   C. joking   D. contradictory

3. The author implies that fungi _____.

    A. are responsible for all the world's rot and decay

    B. cannot live completely apart from other plants

    C. are poisonous to resin-producing plants

    D. can survive indefinitely under favorable conditions

4. The author uses the word dilemma (in the first sentence) to indicate that _____.

    A. the fungus is both helpful and harmful in its effects

    B. fungi are not really plants

    C. the function of chlorophyll is a puzzle to scientists

D. the fungus seems to have its own biological laws
5. Which word best describes the fungus as depicted in the passage ?
A. Diligent　　　　　　　　　　B. Enigmatic
C. Parasitic　　　　　　　　　　D. Slothful

Passage B

　　We sometimes think humans are uniquely vulnerable to anxiety, but stress seems to affect the immune defenses of lower animals too. In one experiment, for example, behavioral immunologist Mark Landenslager, at the University of Denver, gave mild electric shocks to 24 rats. Half the animals could switch off the current by turning a wheel in their enclosure, while the other half could not. The rats in the two groups were paired so that each time one rat turned the wheel it protected both itself and its helpless partner from the shock. Laudenslager found that the immune response was depressed below normal in the helpless rats but not in those that could turn off the electricity. What he has demonstrated, he believes, is that lack of control over an event, not the experience itself, is what weakens the immune system.

　　Other researchers agree. Jay Weiss, a psychologist at Duke University School of Medicine, has shown that animals who are allowed to control unpleasant stimuli don't develop sleep disturbances or changes in brain chemistry typical of stressed rats. But if the animals are confronted with situations they have no control over, they later behave passively when faced with experiences they can control. Such findings reinforce psychologists' suspicions that the experience or perception of helplessness is one of the most harmful factors in depression.

　　One of the most startling examples of how the mind can alter the immune response was discovered by chance. In 1975 psychologist Robert Ader at the University of Rochester School of Medicine conditioned mice to avoid saccharin by simultaneously feeding them the sweetener and injecting them with a drug that while suppressing their immune systems caused stomach upsets. Associating the saccharin with the stomach pains, the mice quickly learned to avoid the sweetener. In order to extinguish this dislike for the sweetener, Ader reexposed the animals to saccharin, this time without the drug, and was astonished to find that those mice that had received the highest amounts of sweetener during their earlier conditioning died. He could only speculate that he had so successfully conditioned the rats that saccharin alone now served to weaken their immune systems enough to kill them.

1. Landenslager's experiment showed that the immune system of those rats who could turn off the electricity _____.

  A. was strengthened   B. was not affected
  C. was altered   D. was weakened

2. According to the passage, the experience of helplessness causes rats to _____.

  A. try to control unpleasant stimuli
  B. turn off the electricity
  C. behave passively in controllable situations
  D. become abnormally suspicious

3. The reason why the mice in Ader's experiment avoided saccharin was that _____.

  A. they disliked its taste
  B. it affected their immune systems
  C. it led to stomach pains
  D. they associated it with stomachaches

4. The passage tells us that the most probable reason for the death of the mice in Ader's experiment was that _____.

  A. they had been weakened psychologically by the saccharin
  B. the sweetener was poisonous to them
  C. their immune systems had been altered by the mind
  D. they had taken too much sweetener during earlier conditioning

5. It can be concluded from the passage that the immune systems of

animals _____.

    A. can be weakened by conditioning

    B. can be suppressed by drug injections

    C. can be affected by frequent doses of saccharin

    D. can be altered by electric shocks

第二单元 生命物质的生物化学
UNIT TWO
BIOCHEMISTRY OF LIVING MATTER

Living matter, or protoplasm, cannot be defined adequately. It differs from lifeless material in possessing the capabilities of growth, repair, and reproduction. These properties may not be apparent at all times in the same degree, but they are present to some extent in all living organisms. Moreover, the life processes go on at comparatively low temperature and with great rapidity, the synthesis of a complex protein molecule such as hemoglobin, for example, apparently requiring only a few seconds. Comparable reactions in the laboratory, even if possible, require high temperatures, often with increased pressure, or else they go on very slowly and quite incompletely. Many reactions of the living cell are of great complexity-intricate interwoven oxidations, disintegrations, and syntheses-in comparison with which the manifold simultaneous operations of an electronic computer are like simple mechanical toys. Some of these marvelous reactions are known and partly understood. Many others are appreciated only because of our awareness of the end products. We must be impressed by the orderly way in which all the chemical activities of the body coordinate. This may be another attribute of living matter, the orderliness of its chemical reactions.

Protoplasm is composed of water, inorganic salts, and organic compounds. Water is a most important compound in tissues and comprises some 75% to 85% of the weight of most cells. The water of the tissues and body fluids is mostly in the free state; i.e., substances may be dissolved in it and it may pass back and forth from blood to tissues, in and out of cells. A small fraction of the water is believed to be bound. In other words, some of the water in hydrophilic colloid systems is combined so that the activity of the water molecules is reduced considerably. Free water varies according to diet and physiologic activity, whereas bound water is a rather constant constituent of the tissues.

Recent studies using deuterated water (D_2O) in dogs have shown that the average water content of the body as a whole is 61% of body weight, with a range of 55% to 67%. The water content of the human body apparently has about the same range, being less than average in fat individuals and somewhat greater in thin persons. The water content of individual tissues also varies considerably.

There are several mechanisms for maintaining and controlling the water content of the tissues. When these go wrong, a number of pathologic states may ensue. Dehydration is a condition not at all uncommon and is likely to have a fatal outcome if not recognized and combatted. Edema is another—a condition in which fluid leaves the bloodstream and accumulates in the tissues. Sometimes what appears to be a minor disturbance results in a major catastrophe.

Water is needed for many and varied reasons. It is the solvent, the agency that enables water-soluble, water-miscible, or emulsifiable substances to be transferred in the body, not only in the blood, which is more than four-fifths water, but also intercellularly and intracellularly. Ionization takes place in water, and ionization is a prerequisite to many biochemical reactions.

In the regulation of body heat, water is most important because of its peculiar physical properties. It possesses high specific heat, i. e., the amount of heat required to raise the temperature of a gram of water 1℃ is much higher than the amount of heat required to raise the temperature of a gram of some other substance 1℃. The specific heat of water is 1. The values for all other common substances are much smaller. This enables the body to store heat effectively without greatly raising its temperature. Water has high heat conductivity. This permits heat to be transferred readily from the interior of the body to the surface. Finally, water possesses high latent heat of evaporation, which causes a great deal of heat to be used in its evaporation and thus cools the surface of the body. These are physical properties useful to the body in the physiologic regulation of body temperature.

At least 60 of the 102 or more elements believed to be present in the universe occur in biologic matter. Only some 20 to 22 of these are found consistently, however, and some are present only in extremely minute amounts. A number of elements occur in living matter as mixtures of the more common form with varying amounts of other forms of the same element. These have slightly different atomic structure and atomic weight from the more common form and are called isotopes. Thus ordinary chlorine, with an atomic weight of 35.453, has been found to be a mixture of two isotopes, the first and more abundant one having an atomic mass of 35 and the second, less abundant one, an atomic mass of 37. Since isotopes in general have the same chemical and biologic properties as the more abundant form, they have proved extremely valuable as tracers in biochemical research.

With the exception of water and small amounts of gases, e. g., oxygen and carbon dioxide, the remaining chemical constituents of living matter consist of organic compounds: carbohydrates, lipids, proteins and many others. The various tissues differ in all of these constituents qualitatively and quantitatively. It is to be expected that a nerve cell will not have the same composition as a salivary gland cell. However, all cells resemble each other chemically to some extent.

生词和词组
New Words and Expressions

protoplasm	原生质；细胞质
synthesis	合成（法）；综合（物）
protein	蛋白质，朊
hemoglobin	血红蛋白
intricate	复杂的，错综的；缠结的；难懂的
interweave	（使）交织；（使）混杂；使紧密结合
disintegration	分解，分裂；衰变，蜕变；崩溃
manifold	多种多样的；多方面的

attribute	属性；特征；品质；把……归因于
hydrophilic	亲水的
colloid	胶体；胶质，胶态
physiologic	生理的；生理学的
deuterate	使氘化
pathologic	病理学的；病态的
ensue	接着发生；结果产生
dehydration	脱水
edema	浮肿；水肿
miscible	能溶和的；易混合的
water-miscible	水混性的
emulsifiable	可乳化的
intercellularly	细胞间
intracellularly	细胞内
ionization	离子化；电离
prerequisite	先决条件，前提；必要条件
conductivity	传导性；传导率
latent	潜在的；潜伏的
isotope	同位素
chlorine	氯
tracer	示踪；示踪物；曳光剂
carbohydrate	碳水化合物；糖类
lipid = lipoid	类脂（化合）物；类脂体
salivary	（分泌）唾液的
specific heat	比热

注解
Notes

1. Many reactions of the living cell are of great complexity...in comparison with which the manifold simultaneous operations of an electronic computer are like simple mechanical toys.

此句中 in comparison with which…意为"与……相比"，其中关系代

词 which 引导的从句做介词的宾语，其先行词是 reactions of the living cell。

2. with a range of 55％ to 67％意为（身体中含水量的）变化范围在（体重的）55％到67％之间。

3. being less ⋯and somewhat greater⋯这是一个现在分词短语作状语的结构，其逻辑主语是该句的主语 the water content of the human body。

4. It is the solvent, the agency that⋯这是一个强调句型，句中的 agency 是 solvent 的同位语。

译文参考
Translation for Reference

生命物质的生物化学

我们无法给生命物质（即原生质）下一个确切的定义。它不同于无生命物质，因为它能生长、修补和繁殖。这些特性可能并不是在任何时候都以同等程度显示出来，然而一切活的有机体或多或少都有这些特性。此外，生命过程是在较低的温度下、十分迅速地延续的，例如，血红蛋白之类的复杂蛋白质分子的合成显然仅需几秒钟。在实验室内作类似的反应，即使办得到，也需要高温，而且往往还要增加压力，否则反应就进行得十分缓慢，并且很不完全。活细胞的许多反应，像错综复杂的氧化、分解和合成，是非常复杂的，与之相比，电子计算机的各种同时进行的繁多运算就好像是简单的机械玩具在活动一样了。这些奇妙的反应之中有一些已为我们所知，并且部分为我们所理解。许多其他的反应仅仅是由于我们知道其最后产物而受到重视。我们肯定有深刻印象的是，机体中各种化学反应的相互协调是有条不紊的。这也许就是生命物质的另一属性，即生命物质化学反应的有条不紊性。

原生质是由水、无机盐和有机化合物所组成的。水是组织中最重要的化合物，占大多数细胞重量的 75％到 85％左右。组织和体液中的水大都是自由状态的，也就是说，物质可以溶解于这种水中，这种水可以从血液到组织、从细胞内到细胞外来回流动。据认为小部分水是结合状态的，换句话说，亲水胶体系统中有一些水是与胶体结合的，因此水分子的活性大为降低。自由水随膳食和生理活动而变化，而结合水是组织中相当恒定的成分。

近来把氘化水（D_2O）用于狗身体内的研究表明，整个身体的平均含水量为体重的61%，含水量的变化范围在体重的55%到67%之间。人体含水量的变化范围显然是与此大致相同的，胖的人身上的含水量比平均含水量低，瘦的人身上的含水量比平均含水量稍高些。单个组织的含水量也大不相同。

维持和控制组织中含水量的机制有数种。当这些机制出了毛病，就会产生一些病理状态。脱水并不是不常见的状态，如果不能识别并加以治疗的话，可能会产生致命的后果。水肿是另一种状态：液体离开血液，储积在组织内。有时，看来好像是轻微的失调，而结果却导致一场大病。

需要水的理由是多种多样的。水是溶剂，它能使水溶性、水混性或可乳化的物质在身体内输送，不仅是在血液内，血液4/5以上是水，而且是在细胞间和细胞内输送。离子化作用是在水中发生的，它是许多生化反应的必要条件。

由于水的特殊物理性质，水在调节体温方面是最重要的。水的比热高，也就是说，1克水温度升高1℃所需的热量比1克其他物质温度升高1℃所需的热量要高得多。水的比热是1，其他各种普通物质的比热数值要小得多，这就使身体能有效地储存热量而不会使体温上升太高。水的导热性很强，这就使热很容易从体内传到体表。最后，水的蒸发潜热很高，这就使大量的热用于水的蒸发上，从而降低了体表的温度。水的这些物理性质对于身体进行体温的生理调节是有用的。

宇宙中存在的102种以上的元素当中，至少有60种出现于生物物质之中。然而，其中只有20种～22种元素是始终存在的，有些元素仅是极其微量地存在，有一些元素是以其普通形式与同一元素的不等量的其他形式的混合体存在于生命物质之中。这些其他形式在原子结构和原子量上与普通形式略有不同，被称为同位素。因此，普通的氯（原子量为35.453）已发现为两种同位素的混合物。第一种含量较高，原子量为35；第二种含量较低，原子量为37。由于一般的同位素和含量较高的同位素具有相同的化学和生物性质，因此业已证明同位素在生化研究方面用作示踪物是极其有价值的。

除了水和少量的气体（例如氧和二氧化碳）之外，生命物质的其余的化学成分是由有机化合物组成的：糖类、脂类、蛋白质和许多其他化合物。各种组织含有的所有这些组分在质和量上是不同的。可以想象，神经细胞不会与唾液腺细胞具有同样的组分。然而，所有细胞在化学上多少是相似的。

练习
Exercises

练习一　视译原课文内容

练习二　阅读理解

Conventional wisdom about conflict seems pretty much cut and dried. Too little conflict breeds apathy and stagnation. Too much conflict leads to divisiveness and hostility. Moderate levels of conflict, however, can spark creativity and motivate people in a healthy and competitive way.

Recent research by Professor Charles R. Schwenk, however, suggests that the optimal level of conflict may be more complex to determine than these simple generalizations. He studied perceptions of conflict among a sample of executives. Some of the executives worked for profit-seeking organizations and others for not-for-profit organizations.

Somewhat surprisingly, Schwenk found that opinions about conflict varied systematically as a function of the type of oxidation. Specifically, managers in not-for-profit organizations strongly believed that conflict was beneficial to their organizations and that it promoted higher quality decision making than night be achieved in the absence of conflict.

Managers of for-profit organizations saw a different picture. They believed that conflict generally was damaging and usually led to poor-quality decision making in their organizations. Schwenk interpreted these results in terms of the criteria for effective decision making suggested by the executives. In the profit-seeking organizations, decision-making effectiveness was most often assessed in financial terms. The executives believed that consensus rather than conflict enhanced financial indicators.

In the not-for-profit organizations, decision-making effectiveness was defined from the perspective of satisfying constituents. Given the complexities and ambiguities associated with satisfying many diverse constituents executives perceived that conflict led to more considered and acceptable decisions.

1. In the eyes of the author, conventional opinion on conflict is _____.

    A. wrong
    B. oversimplified
    C. misleading
    D. unclear

2. Professor Charles R. Schwenk's research shows _____.

    A. the advantages and disadvantages of conflict
    B. the real value of conflict
    C. the difficulty in determining the optimal level of conflict
    D. the complexity of defining the roles of conflict

3. We can learn from Schwenk's research that _____.

    A. a person's view of conflict is influenced by the purpose of his organization
    B. conflict is necessary for managers of for-profit organizations
    C. different people resolve conflicts in different ways
    D. it is impossible for people to avoid conflict

4. The passage suggests that in for-profit organizations, _____.

    A. there is no end of conflict
    B. expression of different opinions is encouraged
    C. decisions must be justifiable
    D. success lies in general arrangement

5. People working in a not-for-profit organization _____.

    A. seem to be difficult to satisfy
    B. are free to express diverse opinions
    C. are less effective in making decisions
    D. find it easier to reach agreement

第三单元 论文摘要
UNIT THREE
ABSTRACTS OF PAPERS

SOIL DEGRADATION IN CHINA:
IMPLICATIONS FOR AGRICULTURAL SUSTAINABILITY,
FOOD SECURITY AND THE ENVIRONMENT
中国土壤退化：
对农业可持续性，食品安全和环境的影响

This dissertation consists of one introduction chapter and three essays, which describe and discuss methods to address three separate but related issues in soil management in China. In my introductory Chapter, I discuss the background for the soil degradation in China and how soil degradation threatens food security, the environment and agricultural sustainability.

本文包括一篇导论和三篇文章，描写和讨论解决中国土壤管理的三个不同却相关的问题。导论部分，笔者讨论了中国土壤退化的背景以及土壤退化如何威胁食品安全、环境以及农业可持续性的问题。

In the first essay in Chapter 2, I develop a dynamic optimization model for soil management and provide implications for the influence of externalities on intertemporal management of soil capital. This chapter contributes to the literature by providing a more comprehensive dynamic optimization model from a social planner's standpoint, who is concerned about agricultural sustainability, environmental quality and food security. A comparison by numerical methods between a public model and a private model implies that optimal soil management path is different for farmers than for social planners when externalities are considered. This implies that it is important to take externalities into account when managing natural capital such as soil. Food security, as a positive externality, and environmental pollution, as a negative externality, are complementing each other. Factors affecting farm profits and externalities also affect the optimal path.

在第一篇文章的第二章，笔者制定了一个土壤管理动态优化模型以及就外部性对土壤资金跨期管理的影响提供了启示。本章通过从关心农业可持续性，环境质量和食品安全的社会规划者角度提供了更全面的动态优化模型，对文献做出了贡献。由公共模型和私人模型之间的数值方法的比较表明，当考虑到外部性时，相比于社会规划者，最佳土壤管理路径对农民来说是不同的。这意味着，当管理自然资本如土壤时，考虑外部性因素是重要的。食品安全，是一个积极的外部性因素，而环境污染，是一个消极的外部性因素，二者相互补充。影响农场利润和外部性的因素也影响最优路径。

In Chapter 3, I propose environment-adjusted profit as a more appropriate tool to measure the costs imposed by environmental regulations than abatement costs from a shadow pricing model. Environment-adjusted profit updates abatement costs by taking farmers' mitigation behavior into account. Both abatement costs and environment-adjusted profit are estimated for over 1,700 cropping systems in the Loess Plateau of China. Furthermore, a regression was used to determine the cropping systems that are most profitable as environmental regulations were imposed. Results show that conservation techniques and mono-crop corn and rotations such as corn-soybean-corn and alfalfa 3 years-corn-millet contribute more to farm profit if environmental regulations were imposed. The conclusions from this chapter can provide farmers and policy-makers alternative choices to balance both economic and environmental goals, rather than planting all land to trees through the Grain for Green program, which was the choice for many in the Loess Plateau.

在第三章中，笔者提出环境调节利润作为一个比影子定价模型中的减排成本更合适的工具来测量环境法规所规定的成本。环境调节利润通过考虑农民的缓减行为更新减排成本。对中国黄土高原1700多种耕作体系的减排成本和环境调节利润进行了估算。此外，用回归分析来确定那些环境法规中规定的最有利可图的耕作制度。结果显示，如果环境法规规定了，保护技术如单一作物（玉米）和轮作（玉米—大豆—玉米，三年苜蓿—玉米—小米），则更有利于农场利润。本章的结论能够为农民和决策者提供替代选择来平衡经济和环境目标，而不是通过退耕还林计划把所有的土地种上树，虽然这是黄土高原上许多人的选择。

In Chapter 4, I update the sustainable value approach by a DEA benchmark and apply it to the cropping systems in the Loess Plateau of China to investigate sustainable value and efficiency as measures of sustainability. The cropping systems that contribute the most to sustainability from the perspective of using all types of capital efficiently are identified by a regression model. Sustainable value and efficiency matrices are created to compare the sustainability between any pair of rotations and conservation techniques. Rotations such as CSC, A3CM and FA5MC are most sustainable. Conservation techniques such as terracing, mulching and furrow-ridging are more sustainable. This chapter contributes the literature in soil science by adding economic perspective in analyzing agronomic techniques.

在第四章中，笔者通过DEA基准更新可持续估价法，并将其应用于中国的黄土高原耕作制度中来调查可持续价值和可持续措施的效率。通过回归模型认定使用所有有效资本类型的最有利于可持续性的耕作体系。建立可持续价值和效率模型来比较任一对轮作和保护技术之间的可持续性。轮作如CSC，A3CM和FA5MC最可持续，而保护技术，诸如建造梯田，覆盖，开沟起垄则较为可持续。本章通过增加经济角度分析农业技术对土壤科学文献做出了贡献。

POTENTIAL ENVIRONMENTAL IMPACTS FROM CROPPING-PATTERN AND LAND-USE CHANGES UNDER THAILANDS's ETHANOL PRODUCTION MANDATE
泰国乙醇生产授权下的种植方式和土地利用变化的潜在环境影响

The primary energy source meeting demand in Thailand is oil, especially in the transportation sector, which has resulted in energy import dependency and environmental impacts (Energy Policy and Planning Office, 2012). To reduce energy import and carbon emission the Thai government has announced a plan, known as "Low Carbon Society" policy that promoted bioenergy use (Ministry of Energy, 2012). The main bioenergy strategy of the Thai government is promotion of ethanol production. Ethanol production targets have been set at 3.0, 6.2, and 9.0 million liters per day, in 2008—2011, 2012—2016, and 2017—2022, respectively (Ministry of Energy, 2012).

在泰国满足需求的主要能源是石油,尤其是在交通运输部门,这造成了对能源进口的依赖性和环境影响(能源政策和规划办公室,2012)。为了减少能源进口和碳排放,泰国政府已经宣布了一项计划,称为"低碳社会"提高生物能源使用政策(能源部,2012)。泰国政府的主要生物能源战略是推广乙醇生产。乙醇生产目标已分别定为2008—2011年:每天3,000,000升,2012—2016:6,200,000升,以及2017—2022:9,000,000升(能源部,2012)。

The main feedstocks for ethanol production in Thailand are cassava and molasses, a byproduct from refining cane sugar. The cultivation areas of these energy crops are thus expected to increase and intensify due to expansion on ethanol production. In 2010, it was estimated that 1.61 million tonnes of cassava and 2.19 million tonnes of molasses could serve as feedstock for ethanol production of 2.25 million liters per day. Based on licensed ethanol plants and the ethanol production target for 2022, demand for cassava and molasses from the Thai ethanol industry would increase up to at least 14.34 and 3.96 million tonnes per year. While the current molasses production could serve this feedstock demand, the enormous increase in demand for cassava would significantly increase land-use for cassava cultivation.

乙醇生产在泰国的主要原料是提炼蔗糖的副产品——木薯和糖蜜。因此这些能源作物的种植面积因乙醇生产扩大而有望增大。据估计,在2010年,161万吨木薯和219万吨糖蜜可作每天生产2,250,000升乙醇的原料。基于许可的乙醇工厂2022年的乙醇生产目标,对泰国乙醇工业中木薯和糖蜜的需求将增加到每年至少14,340,000吨和3,960,000吨。虽然当前糖蜜生产可以满足这个原料需求，但对木薯巨大的需求增加将大大增加木薯种植的土地利用面积。

The ethanol production has been promoted for the purpose of energy security, GHG emission reduction, and economic development. However, it is unclear that the ethanol target of the Thai government is possible in both economic and political terms regardless of the cropping land-use change and thus the environmental impacts. Moreover, the planning, monitoring, and setting suitable cultivation area for ethanol feedstock could help to reduce its negative impact on land use change, deforestation, and biodiversity loss (Scarlat and Dallemand, 2011). This proposed study thus focuses on three interrelated topics: the economic and political feasibility of enacting these mandates; the potential cropping land-use change under realistic scenarios; and the potential environmental impacts of these changes. The objectives for each of these are as follow:

乙醇生产已提升为能源安全、减少温室气体排放和经济发展的目的。然而，目前尚不清楚在不考虑种植土地利用变化和因此造成的环境影响的情况下，泰国政府在经济和政治方面的乙醇目标是否可能。此外，规划、监测和设置适合种植乙醇原料的区域可能有助于减少对土地利用变化、森林砍伐和生物多样性消失（Scarlat和Dallemand，2011年）的负面影响。因此，本研究提出的重点是三个相互关联的主题：制定这些任务的经济和政治上的可行性、现实情况下潜在的种植土地利用变化以及这些变化的潜在环境影响。对于每个变化的目标如下：

To evaluate the current economic and political feasibility to produce nine million liters per day of ethanol. The economic feasibility regards to estimate adequacy of ethanol feedstock crops and cultivate areas as compared to other major competing crops benefit. The political feasibility issues regards the competition of interests among influential parties that play important roles in the Thai energy and agricultural industries, such as the government itself, oil companies, and farmer associations.

评估目前经济和政治上的可行性来生产每天 9,000,000 升乙醇。经济上的可行性方面在于，相比于其他主要竞争作物益处，估计乙醇原料作物和种植面积充足。在政治上的可行性问题涉及在泰国的能源和农业等行业发挥重要作用的有影响力的各方之间的利益竞争，如政府本身、石油企业和农民协会。

To assess on the outcome of cropping land-use change when ethanol target is introduced. The significant increase in ethanol and feedstock demand is expected to dramatically alter crop cultivation areas. Moreover, energy crops and competitive crop prices would also impact on farmers' decision. Thus, individual farmers' economic decision when adopting ethanol feedstock crops to be cultivated instead of other competitive crops will be investigated. Various scenarios cropping land-use change when ethanol mandate is implemented and subsequent will be studied in-depth by using the Multi-criteria Analysis and Geographic Information Systems (GIS).

评估制定乙醇目标时种植的土地利用变化的结果。乙醇和原料需求的显著增加有望戏剧性地改变作物种植面积。此外，能源作物和竞争性作物的价格也会影响农民的决定。因此，将调查当采用乙醇原料作物种植而不是其他竞争性作物时的个体农民经济决策。将利用多准则分析和地理信息系统（GIS）来深入研究当乙醇生产授权执行时种植业土地利用变化的各种方案和后续情况。

To estimate the environmental impacts of Thai ethanol mandate under these various scenarios. Ethanol mandate implementation does not only directly affect GHG reduction, but also effects GHG balance due to cropping land-use change. Other environmental impact such as biodiversity can also be measured. Based on a range of realistic alternative scenarios of cropping land-use change, the range of impacts on several measures of environmental quality will be estimated. The CENTURY model will be used to account soil carbon sequestration as GHG balance. Meanwhile, the nitrous oxide, methane, and biodiversity loss from cropping land-use change are discussed.

估计在各种各样场景下的泰国乙醇政策的环境影响。实现乙醇政策不仅直接影响减排温室气体,而且也影响因耕作土地利用变化所致的温室气体平衡。也可以测量其他环境影响,例如生物多样性。基于一系列的种植土地利用变化的现实替代方案,将对环境质量的几项措施的影响范围进行估算。将用世纪模型来解释土壤碳封存如温室气体平衡,同时,讨论一氧化二氮、甲烷和生物多样性在土地利用变化中的消失。

THE ECONOMIC IMPACTS AND PERFORMANCE OF IRRIGATION IN ZIMBABWE
经济影响与津巴布韦的灌溉实施

This dissertation addresses some issues concerning the performance of irrigation in Zimbabwe. Experts agree that there are many places in the world, Zimbabwe included, where food supply could better match demand through increased production. One way of increasing food production is through the development of water resources for irrigation.

本文设法解决一些有关津巴布韦灌溉实施的问题。专家认为世界上有许多地方,包括津巴布韦在内,可以通过提高产量使食物的供给更好地适应需求。而提高食物产量的方法之一就是开发用于灌溉的水资源。

In this study the benefits of irrigation are estimated for maize production in Zimbabwe. A Marshalian partial equilibrium analysis is used to estimate the net benefits for consumers and producers in maize production. The study concludes that the net benefits are high for a strategy that pursues the development of smallholder irrigation as opposed to one that develops large-scale commercial systems.

在这项研究中,灌溉的好处在津巴布韦的玉米产量上得以估量。用一个 Marshalian 局部均衡分析来估计消费者和生产者在玉米产量上的净收益。研究表明,追求小农灌溉策略的开发净收益高,而非发展大规模商用系统策略的开发净收益高。

Smallholder irrigation performance in Zimbabwe is assessed in terms of water management and financial performance. The conclusion from the study suggests that smallholder irrigation is a viable development alternative. Smallholder irrigation is defined as the formal systems namely the government managed Agritex systems, the farmer managed community systems and the parastatal supported ARDA outgrowers.

津巴布韦的小农灌溉实施由水资源管理和财务业绩来评定。研究结论表明，小农灌溉是一项切实可行的发展选择。小农灌溉定义为正规系统即政府管理 Agritex 系统，农民管理社区系统和半官方的 ARDA 种植者支持。

This study compares the performance of the formal systems themselves and then the formal systems are compared to the informal *bani* system. Because of lack of data the ARDA systems are only included in very few of the analyses. The informal *bani* system is discouraged by the government through statutory instruments.

这项研究先比较正规系统本身实施情况，然后将正规系统与非正规的 bani 系统进行比较。因为缺少数据，所以只有极少数的分析中包括了 ARDA 系统。政府通过法定文件限制非正规巴尼系统。

The main conclusion from the performance of the formal systems is that the farmer managed community system outperforms the government managed system in terms of water management and financial performance. To this extent, the government should utilize the farmer-managed model for further smallholder irrigation development.

关于正规系统性能的主要结论是农民管理社区系统在水资源管理和财务业绩上优于政府管理系统。在这个意义上，政府应该利用农民管理模式进一步发展小农灌溉。

The study also concludes that the informal *bani* system, which is discouraged by the government, can achieve the objectives of irrigation in much the same way, if not better than the formal systems. However, since there is legislation that discourages the utilization of this resource, it is recommended that the government reviews the legislation, in light of contemporary knowledge, in order to allow this apparently valuable resource to contribute to agricultural development in Zimbabwe.

研究还得出结论,由政府限制的非正规 bani 系统,即使不优于正规系统,也能以与正规系统大致相同的方式完成灌溉目标。然而,由于法律限制使用这种资源,人们建议政府根据当代知识审查立法,让这明显宝贵的资源在津巴布韦的农业发展中做出贡献。

USING ECONOMETRICALLY-ESTIMATED IMPORT DEMAND SYSTEMS TO ANALYZE DOMESTIC AND INTERNATIONAL ISSUES: A CASE STUDY OF SAUDI ARABIAN AGRICULTURAL IMPORTS
使用计量经济学估计进口需求系统分析国内和国际问题:以沙特阿拉伯农产品进口为例

With dramatically growing food consumption, combined with the country's climatic conditions and water scarcity, Saudi Arabia depends heavily on food imports to cover the gap between domestic demand and local production. This increasing reliance on imports, along with expected effects of the World Trade Organization (WTO) negotiations on agriculture, and declining domestic production due to changes in domestic policy are the main problems facing the Saudi agricultural sector. This suggests the need to evaluate effects of domestic and international policy change on imports, local production and local demand for key products in the sector.

由于急剧增长的食物消费,加上国家的气候条件和水资源短缺,沙特阿拉伯在很大程度上依赖于食品进口来弥补国内需求和当地产量之间的差距。对进口的依赖不断增加,对世界贸易组织农业谈判的预期效应以及由于国内政策变化所导致的国内产量下降,这些都是沙特农业部门所面临的主要问题。这表明,沙特需要评估国内和国际政策变化对进口的影响,对当地产量和当地对行业内重点产品需求的影响。

The objectives of this research are, first, to empirically analyze the demand for food imports using recent methodological developments in the implementation of demand systems. Then, these estimated demand parameters are utilized to analyze domestic and international policy issues. To attain these objectives, a demand system for imports was estimated using several systems (Rotterdam, AID S, CBS, and NBR), which are all nested alternatives in a general model, and then expenditure and price elasticities were calculated.

此项研究的目的，一是使用完成的需求系统中最新发展的方法论来实证分析对食物进口的需求，二是利用这些估计需求参数来分析国内外政策问题。为了达到这些目标，利用多个一般模式中的嵌套替代系统（Rotterdam，AID S，CBS，and NBR）来估计进口需求系统，然后计算支出和价格弹性。

The CBS model was found to be the best model for analyzing import behavior in this study, and it was used to compute expenditure elasticities, own-price elasticities, and cross price elasticities. The estimated expenditure elasticities imply that if aggregate consumer expenditure allocated to agricultural imports was to increase, the demand for imports would increase. Also, prices show unitary elasticities for six of eight categories.

这项研究发现 CBS 模式是分析进口行为的最佳模式，用以计算支出弹性、自价格弹性和交叉价格弹性。预计支出弹性意味着，如果分配给农产品进口的总消费支出增加，对进口的需求就会增加。价格表明八个类别中的六个为单一弹性。

In order to link domestic and international effects across a variety of alternatives, a restricted regression using the CBS import model was developed to estimate domestic demand elasticities in the presence of alternative supply elasticities. With these matrices of elasticities in hand, a policy analysis framework was developed and simulations of three different issues were made.

为了关联国内外各种替代品的影响，发展使用 CBS 进口模式的约束回归来估计存在于替代供给弹性中的国内需求弹性。随着对这些弹性模型的掌握，发展政策分析框架，对三个不同问题进行模拟。

The first simulation deals with the effect of world price increases from WTO effects, and most import quantities are seen to decline, with the highest decrease for vegetables and fruit (72 percent). There is also an increased incentive for local production, as for example, feed grains expand by 46 percent in response to the WTO price increases. The second simulation deals with the impact on domestic supply due to changes in input prices (water and a subsidy on imported live animals). The results show declining local production because of increasing input prices, with the greatest decline in feed grains (45.5 percent). The reduction in production varies according to input supply elasticities, output supply elasticities, and the local production share in domestic demand.

第一个模拟处理因世界贸易组织影响导致世界价格影响加大问题以及大多数进口数量下降，其中蔬菜水果下降最大（72%）。还有一个增加当地生产的动机，例如，饲料谷物增长46%，以应对WTO的价格上涨。第二个模拟处理因投入价格变化（对进口活动物的水和补贴）导致的国内供给影响问题。结果表明，因为投入价格上升，当地产量下降，其中下降最大的是饲料谷物（45.5%）。产量的减少程度根据投入供给弹性、产出供给弹性和当地生产在国内需求中的份额而变化。

The third simulation shows the effect of increased expenditures on imported goods. The results differ between groups depending on expenditure elasticities and a group's import share in local demand. Once again, feed grains show the highest response (18 percent).

第三次模拟显示了支出增加对进口商品的影响。组与组之间结果不同取决于支出弹性和当地需求中该组的进口份额。再次，饲料谷物显示最高响应（18%）。

RETURNS AND SPILLOVER OF COLORADO AGRICULTURAL RESEARCH COMPUTABLE GENERAL EQUILIBRIUM MODEL FOR LIVESTOCK

科罗拉多农业可计算研究的收益与影响关于牲畜的一般均衡模型

Objectives of this dissertation include study of the implication of investment and returns to livestock research in Colorado as case studies for the Northern Plains Region, description of the components of Colorado regional CGE model of livestock production to analyze direct, indirect and induced impact of investment in livestock research and development of investment recommendations regarding implications of estimation of returns to investment in livestock research in Colorado case study to the regional livestock research policy in the Northern Plains.

本文的目的包括研究对科罗拉多牲畜研究的投资与收益的影响，此研究作为科罗拉多州北部平原地区研究案例，包括描述科罗拉多地区牲畜生产的一般均衡模型的成分来分析对牲畜研究投资的直接、间接及诱发的影响，以及发展关于预计在牲畜研究中的投资收益影响的投资建议，这个牲畜研究属于科罗拉多北部平原的区域性牲畜研究政策中的案例研究。

Starting from initial rate of return of 20%, the estimated ROR was found to be 16%, all direct benefits to the beef as well as all agricultural and non-agricultural sectors included the ROR rises to 25%. When indirect benefits are included, the ROR is estimated to be 28%. When induced effects also added to the benefits of research ROR rises to 35%. Then the CGE model used to simulate 40% ROR to (beef and diary). When all direct benefits (targeted and non-targeted) are included ROR rises to 64%. However, when Indirect benefits included the ROR is estimated to be 76%. When induced effects also added to the benefits of research ROR rises to 104%. Starting from initial rate of return of a 60% ROR to all livestock research the estimated ROR found to be 57%. When all direct benefits (targeted and non-targeted) to the livestock sector as well as all agricultural and non-agricultural sectors are included the ROR rise to 73%. When indirect benefits included the ROR is estimated to be 86%. When induced effects also added to the benefits of research ROR rise to 117%.

从20%的初期收益率开始，预计的收益率将为16%，所有牛肉的直接效益以及所有农业部门包括非农业部门的收益率上升到25%。间接效益包括在内，预计收益率将达到28%。把引发的影响也加入效益研究中，投资收益率上升到35%。一般均衡模型用来模拟40%的投资收益率（牛肉和奶制品）。当所有直接效益（定向的和非定向的）包括在内，投资收益率上升到64%。然而，当非直接收益包括在内，投资收益率预计为76%。当把引发的影响也加入效益研究中，投资收益率上升到104%。从60%的初期收益率开始到所有的牲畜研究，预计收益率将为57%。牲畜部门所有直接效益（定向的和非定向的）和所有农业部门和非农业部门包括在内，投资收益率上升到73%。非直接效益包括在内，预计投资收益率将为86%。引发的影响也加到效益研究上，投资收益率上升到117%。

Investment in livestock research affect returns to factors of production and households income significantly. At 40% ROR to livestock R&D there is about 1% increase in returns to factors of production. The basic livestock research clearly contributes to higher returns to factors of production, and higher income to all household categories. The middle-income group witnesses a slightly higher impact than the low and high-income categories.

对牲畜研究的投资显著影响生产要素和家庭收入的收益。牲畜研发的40%的投资收益率中大约增加1%的生产要素收益。基本牲畜研究显然有利于提高生产要素收益和所有类型家庭的收入。中等收入群体受到的影响略高于低收入和高收入群体。

ECONOMIC ASSESSMENT OF WATER MANAGEMENT IN AGRICULTURE: MANAGING SALINITY AND WATERLOGGING IN THE ARKANSAS RIVER BASIN AND ENVIRONMENTAL WATER SHORTAGES IN THE PLATTE RIVER BASIN
农业水资源管理的经济评估：阿肯色河流域水涝灾害与盐度管理以及普拉特河流域的环境用水短缺

As irrigated agriculture becomes increasingly threatened by both water scarcity and degradation of land and water, the importance of appropriate water management becomes apparent. This dissertation consists of one essay examining the effects of water transfers from agriculture for improving threatened species habitat in the Platte River Basin and two essays addressing irrigation induced waterlogging and salinization in the Arkansas River Basin. Each essay integrates hydrologic modeling into the economic analysis to evaluate the effects of water management alternatives on agricultural production.

灌溉农业越来越受到水资源短缺和水土流失的威胁，适当用水管理的重要性显而易见。本文包括一篇测试调动农业用水来改善普拉特河流域濒危物种栖息地影响的论文，两篇试图解决在阿肯色河流域因灌溉诱发的水涝和盐碱化问题的论文。每一篇论文都将水文模型与经济分析结合来评估水资源管理选择对农业生产的影响。

In the first essay, Discrete Sequential Stochastic Programming (DSSP) is coupled with a basin-wide hydrologic model to estimate the forgone agricultural value associated with water transfers for endangered species habitat restoration. The value of irrigation water in agriculture was estimated for five agriculturally distinct regions of the Platte River Basin. Irrigation water in the upper-most region of the basin was estimated to be of lowest value in agricultural production. Results indicate that although water transfers from agriculture that originate farther upstream result in less water yield at the habitat, they can be more cost effective.

第一篇论文用离散序列随机规划（DSSP）和全流域水文模型来估算为濒危物种栖息地恢复所放弃的农业价值以及调水价值，评估了普拉特河流域的五个农业不同地区的农业灌溉用水的价值。在流域内大多数较高地区，灌溉用水被评估为是农业生产最低值。结果表明，虽然从更远上游区域的农业调水导致栖息地出水量更少，但是这样更有成本效益。

In the second essay, information about current agricultural practices, soil salinity levels, water table depths, and the response functions of crop yields to both waterlogging and soil salinity are used to estimate the current losses associated with waterlogging and salinization. The average forgone profit across the study area was estimated to be $4.3 million annually, or approximately $68/acre per year. This represents the potential of increasing profits by approximately 39% if the effects of waterlogging and soil salinity were removed.

在第二篇中，当前的农业实践信息、土壤盐分含量、水位深度以及土壤盐碱化和水涝情况下作物产量的响应函数都被用来估计当前水涝灾害和盐碱化导致的损失。整个研究区域平均损失的利润估计是每年430万美元，或大概每年每英亩68美元。如果水涝和土壤盐化消除，那么利润可能增长大约39%。

The third essay evaluates several types of alternatives aimed at reducing the impact of waterlogging and soil salinization. The general approach taken is to estimate the costs of inputs required and the commensurate effects on agricultural productivity associated with the changes to soil salinity and water table depth. Each method evaluated was capable of increasing agricultural productivity; however, the associated costs were higher. Although the costs were higher than the direct benefits to agricultural production, significant reductions in salt load to the river were estimated to occur at relatively low costs to society.

第三篇文章评估几种类型的替代方案旨在减少水涝灾害和土壤盐碱化

的影响。所采取的一般方法是估计所需投入的成本以及与土壤盐度与水位深度变化有关的对农业生产力相应影响。每一个评估方法都可以增加农业生产力；然而，相关成本更高。虽然成本比农业生产直接效益要高，但是评估认为相对较低的社会成本能使河流盐负荷显著减少。

OF GOLF AND GRAINS:
THREE ESSAYS ON RESOURCE USE
IN THE NEW AMERICAN WEST
高尔夫和谷物：有关新近美国西部资源使用的三篇文章

The state of Colorado, and indeed much of the Western United States, has experienced rapid population growth and economic development over the past 50 years, transforming this once largely rural region dominated by agriculture and mining into an increasingly urban and suburban population with a diverse economy. The implications of this transformation on the region's natural resource base are numerous and complex. Industries associated with population growth such as golf and house remodeling now have similar gross sales to grain farming and cattle production respectively.

在过去 50 年里，科罗拉多州，实际上大多数美国西部地区，经历了人口的快速增长和经济发展，将以往农业和矿业为主的大部农村地区转变为城市和郊区人口不断增长的多元化经济（区域）。该转变对于该地区自然资源基础的影响多维而复杂。与人口增长有关的产业如高尔夫业和房屋重建业现在分别与谷物业和牲畜生产业有类似的总销售额。

This dissertation examines forces that drive land and water allocations in Colorado and the West. It is found that while limited amounts of land and water are expected to transfer out of agriculture and into industries such as golf over the coming decades, agricultural production will persist and evolve in the West.

本文研究在科罗拉多和其他西部地区，土地和水资源分配的推动力。人们发现在未来几十年里，当把有限的土地和水资源从农业转移到工业如高尔夫时，西部地区农业生产将保留并发展。

REGIONAL AND NATIONAL-SCALE ANALYSIS OF CROPLAND CARBON CYCLING
区域和国家范围分析农田碳循环

Increased global greenhouse gas (GHG) emissions, including carbon dioxide (CO_2), are known to contribute to global warming. Previous research has found carbon (C) sequestration in agricultural soils as a potential way of mitigating atmospheric CO_2 emissions. In the first part of this study, I evaluated the methods used by Annex 1 (developed) countries in inventorying the sources and sinks of agricultural soil GHG emissions. In the second part, I assessed cropland soil C balance and C storage, considering residue C inputs and CO_2 output from soils, at regional and national scale. One of the main components in this study was estimating the crop residue C inputs, using available county-level yield and area data for major US crops during the period 1982—1997. Since the existing annual data reported by the National Agricultural Statistics Service (NASS) have a large number of gaps (missing data), I filled those gaps by using regression analyses with the data from the Census of Agriculture, and a suite of linear mixed effect models that incorporated county level environmental and economic variables. These comprehensive crop databases were then used to estimate residue C inputs and cropland Net Primary Production (NPP). Interannual and spatial variability of residue C inputs were analyzed in relation to changes in production, weather and climate. I also evaluated the potential use of Advanced Very High Resolution Radiometer (AVHRR) Normalized Difference Vegetation Index (NDVI) data in estimating crop aboveground biomass and residue C inputs. Finally I estimated the soil C stocks and annual stock changes in cropland soils (and CO_2 loss due to decomposition), by using the Introductory C Balance Model (ICBM), and evaluated the C storage in soil and overall C balance over the US cropland.

全球温室气体（GHG）包括二氧化碳（CO_2）排放量的增加导致了全

球变暖。先前的研究已经发现农业土壤中碳（C）储存是减少大气中的二氧化碳排放的潜在方式。在本研究的第一部分，笔者评估了附件1发达国家总结农业土壤温室气体排放的来源和下沉所使用的方法。在第二部分，从区域和国家范围，考虑到土壤残留碳投入和二氧化碳排放，笔者评估了农田土壤碳平衡和碳储存。本研究的主要部分之一是估算作物残留碳输入，使用了1982—1997年间美国主要农作物可用的县级产量和面积数据。因为国家农业统计局（NASS）报告的现存年度数据有大量的空白（缺失数据），所以通过回归分析与农业人口普查的数据，和一套结合县级环境和经济变量的线性混合效应模型，笔者填补了这些空白。然后用这些综合作物数据库来估计残留碳输入和农田净生产力（NPP）。对涉及产量变化、天气变化和气候变化的年际和空间变化率的残留碳输入进行了分析。笔者也评估了估算作物地上生物量和残留碳输入的先进高分辨率辐射仪（AVHRR）归一化植被指数（NDVI）数据的潜在应用。最后，通过使用引入的碳平衡模型（ICBM），笔者估算了农田土壤（因分解导致二氧化碳缺失）的土壤碳存量和年度存量变化，评估了土壤碳储存和美国农田的整体碳平衡。

ECONOMIC PERFORMANCE OF SMALLHOLDER EXOTIC DAIRY CATTLE IN THE MARGINAL ZONES OF KENYA
肯尼亚边缘地区外来奶牛的小农经济效益

Kenya's Dairy experts in 1970s argued that the "dry marginal zones" could not meet the requirements of the high performing exotic breeds. They recommended the use of upgraded indigenous breeds which have a lower nutritive requirements and greater adaptability even though their milk supply response capability is relatively low. Yet, smallholder farmers have defied expert advice and have instead shown preference for high exotic grade breeds as a key component of their improved milk production strategies. Does this imply that exotic breeds are more profitable than the indigenous breeds? Experts base their advice on research in the high potential zones, not on marginal zones. The common use of exotic breeds by local farmers in the marginal zones provides an opportunity for an applied research project to compare the performance of these two types of cattle breeds.

1970年，肯尼亚乳业专家认为，"干旱边缘地带"不能满足高性能的外来品种的要求。他们建议使用具有较低的营养需求和更大适应性的升级的本土品种，即使他们牛奶供给反应能力相对较低。然而，小农户农民不听专家建议，反而偏爱使用高级外来品种作为他们提高牛奶产量策略的关键部分。这是否意味着外来品种比本土品种更有利可图？专家的建议是基于在高潜力区的研究而非边缘区。边缘区当地农民对外来品种的普遍使用提供了一个应用项目研究机会来比较这两种奶牛品种效益。

Data on dairy farm operations for the period July 2005 to June 2006 is collected and analysed. Three different methodological approaches are applied to determine the performance of dairy farms: (i) stochastic cost frontier to determine economic efficiency; (ii) translog cost function-input demand systems; and, (iii) translog profit function-input demand systems were used to determine various elasticitis and important details on production systems such as input substitutions and economies of scale.

对2005年7月到2006年6月的奶牛场运行数据进行了收集和分析。运用三种不同的方法确定奶牛场的绩效。一是随机成本边界法来确定经济效率；二是超越对数成本函数输入需求系统；三是使用超越对数利润函数输入需求系统来确定不同弹性和生产系统的重要细节例如输入替换和规模经济。

The results indicate that exotic breeds performed better than indigenous breeds: The large breeds have the lowest cost inefficiency. Overall, the mean cost inefficiency of dairy production in the marginal zones is 27.45%. The resource use inefficiency is significantly explained by institutional and socio-economic factors with varying marginal impacts of the respective variables.

结果表明，外来品种效益高于本土品种：大品种有最低的成本无效率。总的来说，边缘区的乳制品生产的平均成本无效率是27.45%。各自变量不同的边际影响的体制和社会经济因素有力解释了资源使用无效率。

The internal workings of the production and marketing systems show that dairy farmers maximize profits and that there are no constant returns to scale. Most of the dairy inputs are complements and the dairy establishment faces diseconomies of scale. The supply response analysis indicates that institutional and socio-economic factors have much greater elastic impacts on dairy production than the price factors. In sum, a further liberalization of the dairy sector would be beneficial to stimulate supply response.

生产和销售系统的内部工作表明，奶农使利润最大化且没有恒定不变的收益。大部分的乳制品输入是作为补充且乳品打造面临着规模不经济的问题。供应响应分析表明，机构和社会经济因素对乳制品生产比价格因素有更大的弹性影响。总之，乳制品行业的进一步自由化将有利于刺激供给反应。

However, institutional and socio-economic setups such as rod infrastructure, extension services, dairy records, credit and education which have a greater impact are required to enhance resource use efficiency and reinforce the liberalization policy. This implies that judicious investments in institutional and socio-economic factors through enhancement of public expenditure are required to promote market oriented smallholder dairy in the marginal zones.

然而，机构和社会经济计划如加强基础设施、推广服务、乳制品记录、贷款和教育有更大的影响，需要提高资源使用效率并加强政策自由化。这意味着需要通过提高公共支出对机构和社会经济因素进行审慎明智的投资来改善边缘区以市场为导向的小农户乳制品。

WATER USE IN THE WESTERN U. S. : IRRIGATED AGRICULTURE, WATER LEASES, AND PUBLIC PREFERENCES
美国西部的水资源利用：农业灌溉、水源租用以及公众偏好

In the western U. S. , water continues to be reallocated from agricultural to urban uses as a result of rapid population growth and urbanization. However, the negative implications of permanent rural-to-urban water transfers call into question the economic practicality and social acceptability of additional transfers. While some of the short-term economic impacts of permanent water transfers have been estimated, less attention has been given to the longer-term impacts of such transfers. There is also a need to evaluate the economic and social viability of emerging alternatives to permanent water transfers.

在美国西部地区，因人口快速增长和城市化导致了水资源继续从农业用途重新分配到城市用途。然而，永久转移农村用水到城市用水的负面影响对附加转移的经济实用性和社会可接受性提出质疑。虽然估计了永久调水的一些短期经济影响，但是对这种转移的长期影响还是给予了较少的关注。也需要评估永久调水的新兴替代选择的经济和社会可行性。

In addition to assessing the economic contribution of irrigated agriculture, this dissertation assesses the economic and social viability of water transfers and some of their alternatives, from the perspectives of both farmers and urban households. Chapter 1 provides a brief overview of western water law and motivation for my research. Chapter 2 assesses some of the longer-term effects of reduced irrigated acreage on the economic health of western rural counties. First, the relationship between irrigated agriculture and rural economic health is modeled via regression analysis of secondary data. The modeled relationship is then examined for structural breaks to test whether there is a minimum level of irrigated land necessary to sustain the economic health of rural agricultural communities. In Chapter 3, a survey of households in the western U.S. uncovers public perceptions and preferences regarding water use, conservation, and reallocation; current levels of water knowledge; and willingness to pay a fee in support of various water conservation and reallocation programs. In Chapter 4, a survey of irrigators in eastern Colorado is used to estimate a supply curve for leased water and to identify some of the factors that influence farmers' decision to lease their water. Chapter 5 concludes and suggests areas for further study.

除了评估灌溉农业的经济贡献，本文从农民和城镇居民的角度评估调水和一些替代方案的经济和社会可行性。第一章简要概述西方水资源法律

PART TWO POPULAR SCIENCES OF AGRICULTURE AND FORESTRY

和笔者的研究动机。第二章评估灌溉面积减少对于西部农村县域健全经济的一些长期影响。首先，灌溉农业和农村健全经济之间的关系通过辅助数据回归分析建模；然后用建模关系来分析测试结构突变是否有灌溉土地需要的最低水平来维持农村农业社区健全经济。在第三章中，美国西部家庭的一项调查揭示了公众关于用水的认知和偏好、保护和重新分配；当前水资源知识水平；支付费用以支持各种水资源保护和重新分配项目的意愿。在第四章中，用科罗拉多州东部地区灌溉的一项调查来估计一个租赁水的供给曲线，并确定一些影响农民决定出租水资源的因素。第五章总结并提出进一步研究的领域。

The research results will be useful to rural community leaders who are concerned with the evolution of their communities as their resources transition to urban use; urban planners as they consider water supply options; western households as they face the costs of water supply and reallocation programs; policymakers as they consider implementation of water lease markets; and farmers as they consider selling or leasing their water rights.

研究结果将对因资源过渡到城市而担心社区演变的农村社区领导人有用；对考虑供水选项的城市规划者有用；对面临供水成本和重新分配方案的西方家庭有用；对考虑实现水资源租赁市场的政策制定者有用，以及对考虑出售或出租水权的农民有用。

练习
Exercises

练习一　翻译课文原文内容

练习二　将下列汉语段落译成英文

颍上县现代农业发展情况简介

近年来，颍上县按照"抓产业、建基地、扶龙头、求高效、重生态、保安全"的理念，大力实施"结构、规模、合作、转化"四大农业发展战略,以"农业产业化、农村社区化、农民现代化"为工作重心,全力推动现代农业发展,全县农业农村发展了取得一定成效。

——农业发展规划。依据县域自然条件和生态承载能力,规划建设了省级红星现代农业示范区、省级八里河休闲农业与乡村旅游示范区两个核心示范区,启动建设了颍上县农业经济开发区;以及优质粮食、名特优水产、标准化健康畜禽养殖、无公害蔬菜、观光休闲农业和省级颍上农产品加工产业园区等六大优势产业集聚区。

——农业基础设施。加强了现代农业基础设施装备建设,全县旱涝保收标准农田面积达到105.8万亩,占全县耕地总面积67.77%;万亩耕地农业机械总动力达到6 138千瓦,主要农作物耕种收综合机械化水平达到85.87%;设施农业面积达到20万亩,每万人拥有设施农业面积达到1 286亩。

——农业产业发展。全年粮食总产108.9万吨,粮食生产实现连续十二年丰收;全年出栏生猪91.97万头、肉羊32万只、肉牛7.39万头、禽类555万只,肉蛋禽总产达到12.34万吨。全县水产养殖面积达到16.97万亩,总产量达到4.47万吨。全县农产品质量安全抽检总体合格率100%,蔬菜抽检合格率达100%,水产品抽检合格率达100%。农业产业化龙头企业年加工能力达到156万吨,年加工产值达142.91亿元。

——农业经管机制。全县已组建各类农民合作社1 083个,参加合作社农户比例达到57.6%;全县已流转土地107.4万亩,占耕地总面积的68.8%,已建成规模化、标准化畜禽水产养殖场820家,全县畜禽规模养殖化率达到72%,水产标准化健康规模养殖化率达到73%。农业产业化龙头企业已建立原料生产基地72.5万亩,带动农户21.8万户,占全县农户总数的61.08%。

——农业功能拓展。依托自然优势,积极拓展农业生态保护、休闲观光、文化传承等新型功能,全县休闲观光农业经营单位153家,从业人员达2.46万人。八里河公园被国家旅游局授予5A级风景区;双集茶文化街、焦岗湖渔家乐等7家单位被授予"安徽省乡村旅游100佳精品点"称号,全县被省农委、省旅游局联合授予"全省休闲农业与乡村旅游示范县"称号。2014年全年共接待游客313万人,营业收入达5.54亿元,利税7 500万元,同比分别增长36.7%和33.5%,荣获"全国休闲农业与乡村旅游示范县"称号。

PART TWO POPULAR SCIENCES OF AGRICULTURE AND FORESTRY

——农业科技应用。与安徽农业大学新农村发展研究院合作共建的颍上县草牧业（肉羊）特色试验站是以发展壮大颍上县草牧业、促进农村一二三产业融合为目标，通过资源整合、共建共享等途径，探索建立校县紧密结合、教科推动多位一体的新型大学农业推广服务模式，是集人才培养、技术创新和科技推广、创业孵化四大功能于一体的现代农业特色产业试验站。

译文参考
Translation for Reference

Introduction to the Modern Agricultural Development of Yingshang County

In recent years, according to the concept of "developing industry, building base, helping the big enterprises, pursuing high efficiency, emphasizing ecosystem and ensuring safety", Yingshang county implemented vigorously the four development strategies of agriculture, including "structure, scope, cooperation and conversion". With the focus on "industrialized agriculture, community oriented rural areas and modernized peasants", Yingshang county spared no efforts in promoting the development of modern agriculture, and has achieved a lot.

——Planning for agricultural development. Based on the natural conditions of this county and the bearing ability of ecosystem, Yingshang county has established two core demonstration zones on provincial level: Hongxing demonstration area for modern agriculture and Balihe demonstration area for leisure agriculture and rural tourism. It started to build Yingshang county agriculture economy developing district. Meanwhile, a gathering area characterized by six advantageous industries has been established, which includes high-quality food, famous, special and good aquatic product, standardized livestock and birds raise, green vegetables, sightseeing and leisure agriculture, and Yingshang agricultural product processing industry park on the provincial level.

——The infrastructure of agriculture. Yingshang county has strengthened the build-up of the infrastructure equipment of modern agriculture. The standard farmland that guarantees harvest regardless of drought or waterlogging is up to 1,058 thousand mu, accounting for 67.77% of the total arable farmland. The general agriculture machinery motive power of ten thousand high-quality farmland is up to 6,138 kilowatt. The comprehensive mechanic level of the main crops for ploughing, sowing and reaping is up to 85.87%. Facility agricultural area is up to 200 thousand mu, 1,286 mu possessed by per 10 thousand people.

——The development of agriculture industry. The yearly gross grain yield is 1.089 million tons; the harvest of grain was realized in the past 12 years successively; the output pigs are 919.7 thousand, sheep 320 thousand, cows 73.9 thousand, fowls 5.55 million, and the total production of meat, eggs and fowls is up to 123.4 thousand tons; the gross area for aquaculture in this county reaches 169.7 thousand mu; the gross yield is up to 44.7 thousand tons. The percent of pass of random inspection for agriculture products' quality safety is 100%; the percent of pass of random inspection for vegetables is 100%; the percent of pass of random inspection for aquatic products is 100%. The leading enterprise of industrialized agriculture is capable of processing 1.56 million products, and the yearly output value is 14.291 billion yuan.

——The mechanism of agricultural operation and management. 1,083 different peasants cooperation associations have been founded and 57.6% of the peasants have become members of them; The land that has been transferred is up to 1.074 million mu, accounting for 68.8% of the gross cultivatable land. There are 820 mass scale and standardized raising farms for livestocks, fowls and aquatic products; the mass scope raising rate has reached 73%. The landing enterprise of industrialized agriculture has established 725 thousand mu production base of crude materials, leading and driving 218 thousand rural families, accounting for 61.08% of the total rural families in the whole county.

——The expansion of agricultural function. Based on nature advantages, this county has expanded actively such new functions as protection of agro-ecology, leisure sightseeing and cultural inheritage, and there are 153 units getting involved in the business of leisure sightseeing, and 24.6 thousand people get employed in this field. Balihe Park is granted as 5A level scenic spot by National Tourism Administration; Shuangji tea culture street, Jiaoganghu leisure fishing and other five units were granted as "Top 100 high-quality units for Anhui rural tourism". The county was jointly granted as "The demonstration county of the provincial leisure agriculture and rural tourism" by the Provincial Agriculture Commission and Tourism Administration. In 2014, 3.13 million people visited Yingshang county, bringing 554 million yuan operating income, and 75 million yuan profit and tax, which were respectively increased by 36.7% and 33.5% year-on-year. The county was honored "the National Demonstration County for Leisure Agriculture and Rural Tourism".

——The application of agricultural science and technology. In order to develop and strengthen the grass and animal husbandry, promote the mixing of the three industries in rural areas, Yingshang county has cooperated with Research Institute for New Rural Deveolopment, AAU to establish Yingshang Grass and Animal Husbandry Characteristic Experiment Station. By integrating resources, joint building and sharing, a new all-in-one university agricultural extension service mode has been explored to closely connect university and county, promoting development comprehensively based on education and science. It is a characteristic industry experiment station which integrates the four functions of talents cultivation, technology innovation, science and technology popularization and business incubation.

第四单元　中国农业推广体系的演变与发展（节选）
UNIT FOUR
EVOLUTION AND DEVELOPMENT
OF AGRICULTURAL EXTENSION SYSTEM IN CHINA (EXCERPTED)

农业是人类赖以生存发展的基础。中国人口众多，农业的基础性地位更加突出。随着人类文明进步和经济社会的不断发展，新的技术和生产经营方式不断运用到农业中去，农业技术推广及体系建设对促进农业发展中有着极其重要的作用。农业技术的推广应用是一项复杂的社会系统工程，它不仅包括技术产生系统，技术推广系统和技术消化系统，还包括农业教育与培训、政府与市场职能分工等其他的社会支持系统。这些系统都受到历史和文化要素的深刻影响，尤其是农业推广体制的形成和发展，总是受到技术供给方和技术需求方之间存在的历史和文化因素的深刻影响。中国具有悠久的历史，农业技术推广体系几千年来也在不断演变和发展，分析我国农业科技推广体系的历史演变过程，以及农业推广体系发展变化的规律，对完善中国农业推广体制有着积极的意义。

Agriculture is the basis of human beings' survival and development, and its foundation in China becomes extraordinarily prominent because of China's large population. With the development of human civilization and economic society, new technology and production operations have been successively applied to agriculture, and agricultural technology extension and system construction are extremely important in facilitating the agricultural development. Agro-technology popularization and application is a complex social project, which includes not only the technology creation system, technical extension system and digestive system, but also other social support systems including the agricultural education and training, functional division of government and market, etc. These systems are subject to the profound influence of historical and cultural factors, particularly the formation and development of agricultural extension system are always deeply influenced by historical and cultural factors existing between technical supply side and demand side. China enjoys a long history and its agricultural technology extension system has been evolving and developing for thousands of years. Analysis of historical evolution of China's agricultural technology extension system as well as its regularities of development of agricultural extension system is significant to perfect agricultural extension system in China.

中国农业推广体系的历史演变
Evolution of Agricultural Extension System in China

在中国远古时代，农业生产技术主要通过部落内部，人们共同劳动并集体生活，以观察、模仿、言传身教等方式世代传承；同时也通过部落间的迁徙、接触，以相互影响、效仿、传播的方式进行波浪式的扩散，农业知识与技术经验的传播扩散十分缓慢。到4000年前尧舜时代的奴隶社会，农业生产技术的传承才由自发传播转向自觉推广，并逐步形成行政推广体制。

In ancient times of China, agricultural production techniques were passed on generation after generation primarily in common labor and communal life within tribes by observing and imitating and teaching by precept and example. Meanwhile, the wave-like popularization happened during migration, interaction and imitation among tribes. Dissemination and spread of agricultural knowledge and technical experience were very slow. In the slave society of Yao and Shun Times 4,000 years ago, inheritance of agricultural production technology changed from the spontaneous spread to conscious popularization, and the administrative extension system gradually took shape.

随着人类社会文明程度的不断提高，农业生产力发生了巨大变化，依靠口传身授和简单的文字歌谣的形式来传播和推广农业技术已远不能满足农业生产发展的需求，进入封建社会后，记载农业技术的书籍大量涌现。

数据表明，中国古代农书达500多部，至今仍收藏的有300多部，中国也是世界上拥有农业典籍最多的国家。这些农业典籍是进行农业技术推广传播的载体，也标志着农业技术推广传播由原始的口传身授、歌谣转变为书籍传播，提高了农业技术推广的科学性和准确性。

With the improvement of civilization of human society, agricultural productivity has been changing dramatically. Agro-technology dissemination and popularization in the form of oral imparting with physical instruction and simple letter ballads were far from matching the development of agricultural production. In the feudal society, there emerged in large quantities the books of recording agricultural technology. The data show that Chinese ancient agricultural books are more than 500, still more than 300 are in present collections. China is also the country having the most ancient agricultural classical works in the world. All these works are the vector for spreading agricultural technology, which indicates that agricultural technology extension has been turning from the original dissemination through oral imparting with physical instruction and folk songs to book transmission, improving scientificity and precision of agro-technology popularization.

中国封建社会，历代封建君王为了发展农业经济，沿袭推行劝农政策，实行教育与行政相结合的方针，对农民进行劝导、教育和督导，达到推广农业技术、提高农业生产的目的。如在汉朝，国家实行"劝农"制度，强调以行政手段鼓励农桑，以贯彻君王农本治国政策。自汉朝初期开始，从中央到地方确立劝农官制以后，"劝农"制度历代沿袭。在长期的封建社会中，农业推广活动带有了浓厚的官办色彩和技术、技艺推广特征。这种既强调教育在农业推广中的作用，又用农业行政机构来推行劝农政策的做法，逐渐形成了我国教育与行政措施相结合的农业推广传统。

In the history of Chinese feudal society, feudal monarchs, for the development of agricultural economy, followed and implemented agro-persuasion policies, combining education with the administration, implementing persuasion, education and supervision to the farmers to achieve the objective of promoting agricultural technology and improving agricultural production. As in the Han Dynasty, the nation carried out "Agro-persuasion" system, putting emphasis on administrative measures to encourage farming and sericulture, implementing the monarch's rule of agriculture-oriented policies. Since the beginning of the Han Dynasty, after official Agro-persuasion system was established from central to local administration, the "Agro-persuasion" system passed on through the ages. In the long-term feudal society, the agricultural extension activities featured by official operation plus technology as well as skills promotion. This mode both stressing the role of education in agricultural extension, and utilizing agricultural administration to carry out Agro-persuasion policy, has been gradually evolving into China's agricultural extension culture with combination of education and administrative measures.

更多内容参见二维码

（译者：李清，程逸松，王驰，张小红）

练习

Exercises

练习一　翻译课文内容，做笔译和视译练习

练习二　将下列段落译成英语

安徽农业大学新农村发展研究院
颍上特色试验站简介

"安徽农业大学新农村发展研究院颍上县农区草牧业试验站"以落实创新、协调、绿色、开放、共享发展理念为主线，以发展壮大颍上县草牧业、促进农村一二三产业融合为目标，以形成农区草牧业持续发展的技术体系为重点，以体制机制创新为动力，通过资源整合、共建共享等途径，探索建立校县紧密结合、教科推动多位一体的新型大学农业推广服务模式，为推动和引领颍上县农区草牧业的健康、快速、持续发展和现代生态农业产业化提供人才技术支撑，为构建新型农业社会服务体系探索路径、积累经验。是集科技创新、技术服务、示范带头和人才培养四大功能于一体的现代农业特色产业试验站。

1. 科技创新功能

通过建立"草-羊-土"平衡的草牧业（肉羊）标准化生产体系，在试验站建成一个300亩优质人工混播牧草地的示范基地，研究划区轮牧和农作物秸秆加精料补饲养殖技术，探索出一套种养结合的草牧业生产技术，综合利用滩涂地、低洼地、低产田等，开展草牧业产业化生产。

2. 技术服务功能

通过开展技术培训、技术示范、技术指导、技术服务等工作，培养一批发展理念先进、示范效应明显且具有一定规模的专业化、标准化的草牧业示范和展示基地，将"草牧业特色试验站"建设成为颍上县草牧业先进技术的辐射中心。

3. 示范带动功能

"草-羊-土"平衡的草牧业标准化生产体系的建立，是集种植人工草地、生态养殖、农作物秸秆饲料化利用、羊粪便还田等各项技术于一体，对颍上县2万亩低洼低产地的利用起到示范作用，可以有效利用周边8万亩土地小麦秸秆、大豆秸秆和其他农副产品。同时，带动周边区域少用或不用化肥、农药和杀虫剂，不断改善该区土壤、周边水系和大气环境的质量，从而实现农业生产与环境保护协调可持续发展。

4. 人才培养功能

试验站为安徽农业大学学生和颍上县新型农业经营主体提供机会，提高其实践能力、生产能力、经营能力和就业技能，"草牧业特色试验站"已成为大学生实习实训基地和新型草牧业经营主体培养培训基地。

PART TWO   POPULAR SCIENCES OF AGRICULTURE AND FORESTRY

译文参考
Translation for Reference

The Profile of Yingshang Characteristics Experiment Station, Research Institute for New Rural Development, AAU

"Yingshang County Agricultural Area Grass and Animal Husbandry Experiment Station, Research Institute for New Rural Development, AAU", through resource integration, co-construction and co-sharing and other means, explores the establishment of a new all-in-one mode of university agricultural extension in which universities and counties are closely connected and teaching and science are promoting with implementation of innovation, coordination, green, opening, co-sharing development ideas as the main line, development and expansion of Yingshang County grass and animal husbandry and promotion of primary, secondary and tertiary industrial convergence as the target, formation of the technical system of grass and animal husbandry sustainable development of the rural areas as the key point, innovation of institutions and mechanisms as the dynamic, to provide technical and personnel support to promote and lead the healthy, rapid and sustainable development of grass and animal husbandry and modern eco-agricultural industrialization in Yingshang County Agricultural area, and to explore paths and gain experience to build a new social service system of agriculture. The modern agricultural characteristic industrial experiment station integrates four functions—science and technology innovation, technical services, demonstration lead and personnel training into a whole.

1. Scientific and technological innovation function

A 300-mu demonstration base of high quality artificial mixed seeding pasture has been completed in the station through the establishment of a balance of grass livestock (sheep) standardization of production system of "grass-sheep-soil". Research has been done on rotation grazing, and Crop Straw and concentrate of supplementary feeding breeding technology. A combination of planting and breeding technology has been sought out in grass and animal husbandry, comprehensively utilizing tidal flat, low-lying land, low production field to carry out the grass and animal husbandry industrialized production.

2. The technical service function

A number of specialized and standardized grass and animal husbandry demonstration and presentation bases will be cultivated with advanced development concepts, evident demonstration effect and a certain scale to build "grass and animal husbandry characteristics experiment station" as the radiation center for advanced technology of grass and animal husbandry in Yingshang county through technical training, technical demonstration, technical guidance, technical services and etc.

3. The demonstration and leading function

The establishment of a balanced grass livestock (sheep) standardization production system of "grass-sheep-soil" is an integration of artificial grass planting, ecological farming, feed utilization of crop straw, sheep manure to soil, and other technologies. It serves as a model for the use of 20,000 mu low-lying land low-yield land in Yingshang County. And wheat straw, soybean straw and other agricultural and sideline products in the surrounding 80,000 mu can be effectively used. Meanwhile, the surrounding area can be led to use less or no chemical fertilizers, pesticides and insecticides, the quality of soil, water and atmospheric environment around the region Continuously improved, in order to achieve balanced and sustainable development of agricultural production and environmental protection.

4. The personnel training function

The experiment station provides an opportunity for the students in AAU and of new agricultural management identities in Yingshang County, improving their practice ability, production capacity, management capacity and employability skills. "Grass and animal husbandry characteristic experiment station" has become the internship training base for university students and the cultivating and training base for new grass and animal husbandry management entities.

第三部分　国际交流

PART THREE　INTERNATIONAL EXCHANGE

第一单元 科罗拉多州前沿区域农业价值链创新集群的出现（节选）

UNIT ONE
THE EMERGENCE OF AN INNOVATION CLUSTER IN THE AGRICULTURAL VALUE CHAIN ALONG COLORADO'S FRONT RANGE (EXCEPTED)

Colorado has long embraced agriculture as central to its economy and innovation as an essential driver of economic growth. These two—agriculture and innovation—have been converging in Colorado for some time now, and the pace is picking up.

科罗拉多州长期以来一直将农业作为其经济的核心，并将创新视为经济增长的重要驱动力。这两种农业和创新的结合在科罗拉多州已经有一段时间了，而且速度正在加快。

Innovators in an increasingly integrated Agriculture-Water-Food-Beverage-Bioenergy innovation ecosystem are gathering and growing along Colorado's Front Range, creating next-generation technologies and business models to nourish, refresh, and energize the world.

农业-水-食品-饮料-生物能源创新生态系统日益一体化，其中的创新者们正聚集并沿着科罗拉多州的前沿区域不断发展壮大，创造下一代技术和商业模式来滋养、刷新和激励世界。

The value chain of agriculture
农业价值链

Understanding the full scope and structure of the value chain of agriculture is crucial to understanding the scope of overlapping interests and thus the potential scope for innovation-led clustering dynamics. While the value chain is centered on crop and livestock production, it includes much more.

理解农业价值链的整体范围和结构,对于理解利益重叠的范围,进而理解创新驱动集群动力的潜在范围至关重要。虽然价值链以农作物和牲畜生产为中心,但它包含的内容要多得多。

According to a 2013 study by Colorado State University, the value chain of agriculture can be understood to encompass the entire flow of inputs and outputs that enables agricultural enterprises to realize the value of their unique capital base by meeting the needs of final consumers. The unique capital base of farms and ranches consists of natural capital(land and water), physical capital(equipment, livestock and crop inventories), specialized human capital, and financial capital owned and employed by farm and ranch operations. In 2011 the supply of agricultural inputs—such as seeds, fertilizers, feed supplements, and fuel—by Colorado agribusinesses contributed $2 billion to the state economy. Colorado farms and ranches produced more than $8 billion worth of crop and livestock harvests. Economic activity utilizing those harvests—including commodity marketing, processing, and food and beverage manufacturing—was over $15 billion. The total retail value of products sold in Colorado that derive from agricultural production was over $28 billion. Ultimately, any economic activity derives its value from what final consumers are willing to pay for its contribution to their wellbeing. Altogether, the value chain of Colorado agriculture involves more than 200 separate industry subsectors, as categorized by official NAICS business classification codes.

根据科罗拉多州立大学(Colorado State university)2013年的一项研究,农业价值链可以理解为包含整个投入和产出的流动,使农业企业能够通过满足最终消费者的需求来实现其独特资本基础的价值。农场和牧场独特的资本基础包括自然资本(土地和水)、实物资本(设备、牲畜和作物库存)、专业人力资本以及农场和牧场经营所拥有和使用的金融资本。2011年,科罗拉多州农业企业提供的种子、化肥、饲料添加剂和燃料等农业投入为该州经济贡献了20亿美元。科罗拉多州的农场和牧场生产了价值80多亿美元的农作物和牲畜。利用这些收获的经济活动——包括商品销售、加工和食品饮料制造——超过了 150 亿美元。科罗拉多州农产品零售总额超过 280 亿美元。最终，任何经济活动的价值都源于最终消费者愿意为其对自身福利的贡献付出多少。按照北美产业分类体系（NAICS）官方商业分类代码的分类，科罗拉多州农业价值链涉及 200 多个独立的工业子部门。

Private sector R&D expenditures for the agriculture and food value chain was almost $20 billion worldwide in 2007, according to a recent analysis by the USDA Economic Research Service. In high income countries, such as the U.S., spending on agricultural and food R&D is almost evenly split between the public and private sectors. In evaluating private sector R&D, the USDA study considers innovation in nine industry sectors along the value chain: (1) crop genetic improvement, (2) crop protection chemicals, (3) fertilizers, (4) farm machinery, (5) animal health, (6) animal genetic improvement, (7) animal nutrition, (8) food and beverage manufacturing, and (9) biofuel production. The study finds that R&D trends have been uneven across these nine sectors. R&D has been growing most quickly in crop genetics, farm machinery, and food manufacturing, but has been declining in crop protection chemicals and animal nutrition. The USDA study also finds that private sector R&D in each of the nine sectors has come to be quite concentrated into just a few large multinational companies. This global concentration has significant implications for regional efforts in agricultural R&D.

根据美国农业部经济研究服务部（USDA Economic Research service）最近的一项分析，2007 年，全球农业和食品价值链的私营部门研发支出接近 200 亿美元。在高收入国家如美国，公共部门和私营部门之间在农业和食品研发上的支出几乎平分秋色。在评估私营部门研发方面，美国农业部研究认为创新在九个行业沿着价值链进行：（1）作物遗传改良；（2）作物保护的化学物质；（3）肥料；（4）农业机械；（5）动物卫生；（6）动物遗传改良；（7）动物营养；（8）食品和饮料制造；（9）生物燃料生产。研究发现，这九个行业的研发趋势并不均衡。在作物遗传学、农业机械和食品制造方面，研发一直增长最快，但在作物保护化学品和动物营养方面却在下降。美国农业部的研究还发现，这九个行业的私营部门研发都已相当集中于少数几家大型跨国公司。这种全球集中对农业研发的区域努力具有重大影响。

An inventory of Colorado innovators in the agricultural value chain
一份科罗拉多州农业价值链创新者的清单

Who are the innovators in the agricultural value chain in Colorado? And what are the main technologies or industry sectors in which they are innovating? An inventory was taken of all private sector firms and public sector organizations engaged in innovation, based on (1) those companies and organizations generating the publications and patents identified in the landscape analysis, and (2) referrals from industry associations, networking events, interviews, news accounts, and other expert sources. The inventory includes 550 innovators, of which 460 are private-sector companies and 90 are public-sector (academic, non-profit, and government) organizations.

谁是科罗拉多州农业价值链的创新者？他们创新的主要技术或行业是什么？对所有从事创新的私营部门公司和公共部门组织进行了清查，其依据是（1）产生景观分析中确定的出版物和专利的公司和组织，以及（2）来自行业协会、网络活动、采访、新闻账户和其他专家来源的推荐。这份清单包括550家创新者，其中460家是私营企业，90家是公共部门（学术、非营利和政府）组织。

There appears to be a critical mass of innovating organizations active in Colorado within each of a dozen categories:

在科罗拉多州，一大批创新组织活跃在以下十几个类别：

1. Innovators in water technology, infrastructure, analytics, and management
2. Innovators in soil fertility and pest control
3. Innovators in plant genetics and new crop varieties
4. Innovators in animal health, nutrition, and herd management
5. Innovators in agricultural information systems
6. Innovators in sensors, testing, and analytics for product quality

and biosafety
    7. Innovators in bioenergy
    8. Innovators in commodity processing and food manufacturing
    9. Innovators in dairy production and dairy product manufacturing
    10. Innovators in beer, wine & spirits production and marketing
    11. Innovators in natural, organic, and local foods and marketing
    12. Innovators in "Fast & Fresh" food service
    13. Innovators in other emergent subsectors

    1. 水技术、基础设施、分析和管理方面的创新者
    2. 土壤肥力和病虫害防治方面的创新者
    3. 植物遗传学和新作物品种的创新者
    4. 动物健康、营养和畜牧管理方面的创新者
    5. 农业信息系统的创新者
    6. 传感器、测试、产品质量和生物安全分析方面的创新者
    7. 生物能源的创新者
    8. 产品加工和食品加工领域的创新者
    9. 乳制品生产和乳制品加工的创新者
    10. 啤酒、葡萄酒和烈性酒生产和营销的创新者
    11. 自然、有机、本地食品和营销方面的创新者
    12. "快速新鲜"食品服务的创新者
    13. 其他新兴子领域的创新者

Next steps
后续步骤

0. Based upon this analysis, several next steps are recommended for cultivating and capitalizing upon this economic growth opportunity.

0. 在此基础上,提出了培育和利用这一经济增长机遇的几个步骤。

As a prerequisite, realize the economic significance and technological sophistication of innovation activities occurring in the agricultural and food value chain. The economic significance of introducing game-changing innovations within agriculture, food, water, and bioenergy present real economic opportunity for Colorado.

作为前提，认识农业和食品价值链创新活动的经济意义和技术复杂性。在农业、食品、水和生物能源领域引入改变游戏规则之创新的经济意义为科罗拉多州带来了实实在在的经济机遇。

1. First and foremost, develop and attract talent.

Talent is identified, repeatedly, as the most important factor driving growth of an innovation cluster. The availability of skills is the factor most commonly cited by the executives interviewed for this study. Talent can be attracted to Colorado from other states based on the high quality of life. To develop talent, it falls primarily to universities to supply the kind of high-quality professionals needed in the sciences, engineering, management, law, and finance to really drive the growth of an innovation cluster. For those in the farming and ranching community, there is opportunity for younger generations coming off the farm to combine their knowledge of agriculture with specialized skills in science, engineering, or business.

1. 首先，培养和吸引人才。

人才被反复确认为推动创新集群增长的最重要因素。技能的可用性是这项研究中受访高管最常提到的因素。科罗拉多州的高生活质量可以吸引其他州的人才。要培养人才，大学首先要提供科学、工程、管理、法律和金融领域所需的高质量专业人才，以真正推动创新集群的增长。对于那些在农业和牧场社区的人来说，从农场出来的年轻一代有机会将他们的农业知识与科学、工程或商业方面的专业技能结合起来。

2. Identify and support existing activities, and connect existing companies.

There is already much going on that has arisen in response to market forces and thus has real market potential. Growth of a cluster needs mechanisms to facilitate mixing and the spawning of collaborations. State government and the universities are in an excellent position to invite private sector innovators into networking events and thereby into deeper discussions.

2. 识别和支持现有的活动,并联系现有的公司。

在应对市场力量的过程中,很多事情已经在进展,因此具有真正的市场潜力。集群的增长需要促进混合和协作的生成机制。州政府和大学邀请私营部门创新者参加网络活动,从而在进行更深入的讨论方面处于有利地位。

3. Exercise tolerance of different points of view.

Innovation is, by its very definition, a challenging of the status quo, and it requires a willingness to question how things are done. As a state, Colorado has recently been at the center of national debates, such as labelling of genetically modified organisms or cultivation of industrial hemp. Simply taking sides and defending ones interests is not helpful. Innovation requires listening to other's concerns, respecting others' intellectual and emotional responses to issues, and seeking common ground wherein solutions may lie.

3. 锻炼对不同观点的包容能力。

创新,就其本身的定义而言,就是对现状的挑战,它需要一种质疑事情如何进行的意愿。作为一个州,科罗拉多最近一直处于全国辩论的中心,比如转基因生物的标签或工业大麻的种植。仅仅站在某一方,维护自己的利益是无济于事的。创新需要倾听他人的担忧,尊重他人对问题的智力和情感反应,并寻求解决方案的共同基础。

4. Coordinate vertically, to pilot locally, then sell globally.

Given the complexity of the value chain, vertical coordination is required for piloting many new technologies. The necessary upstream and downstream partners can be found in the Front Range. And, the region's market is large enough to grow within, before seeking to expand nationally and even globally.

4. 垂直协调,在当地试点,然后在全球销售。

考虑到价值链的复杂性,许多新技术的试验需要纵向协调。在前沿区域内可以找到必要的上下游合作伙伴。而且,该地区的市场规模足以在内部实现增长,之后再寻求在全国乃至全球扩张。

5. Develop financing mechanisms to assure access to risk capital.

There may be new opportunities for agricultural innovation by creating financing mechanisms that bring together the market knowledge of agriculturalists with the risk capital expertise of venture investors.

5. 发展融资机制，确保获得风险资本。

通过创建融资机制，将农学家的市场知识与风险投资者的风险资本专长结合起来，这样可能会为农业创新带来新的机遇。

6. Take the long view.

The cultivation of an innovation cluster is a long term effort, measured in decades. By some measures, innovation in the agricultural and food value chain has been mounting already in Colorado for at least two decades. Success may require another decade of dedicated effort.

6. 目光放远。

培育创新集群是一项长期的努力，需要几十年的时间来衡量。从某些方面来看，科罗拉多的农业和食品价值链的创新在过去至少 20 年里一直在增长。成功可能还需要十年期的不懈努力。

Colorado has long embraced agriculture as central to its economy and its Western way of life. Colorado has also long embraced innovation as an essential driver of economic growth and part of its pioneer culture. Yet, Colorado has not always put these two together, at least within the public imagination. But, these two—innovation and agriculture—have been converging in Colorado for some time now, and the pace is picking up.

科罗拉多州长期以来一直把农业作为其经济和西部生活方式的中心。科罗拉多州长期以来也一直将创新视为经济增长的重要驱动力，并将其作为先锋文化的一部分。然而，科罗拉多并不总是把这两者放在一起，至少在公众的想象中是这样。但是，这两项创新和农业已经在科罗拉多州融合了一段时间，而且步伐正在加快。

A number of geographic, demographic, and economic factors are driving greater investment and engagement in innovation in the agricultural and food system in Colorado. These forces are particularly strong in the Denver metro region and the northern Front Range. A number of companies and investors have taken note of these factors—or have happily stumbled upon them—as they have started new ventures, invested in existing businesses, or moved operations into this region.

许多地理、人口和经济因素正在推动科罗拉多州农业和粮食系统创新方面的更多投资和参与。这些力量在丹佛都会区和北部前沿区域尤其强大。许多公司和投资者已经注意到了这些因素——或者很高兴地偶然发现了它们——因为他们已经开始了新的风险投资，投资于现有的业务，或者将业务转移到了这个地区。

These factors include the quality of the region's talent pool, the strength of public research institutions, the sheer amount of primary agricultural production, the proximity of that agricultural production to other economic activities, and the size of the local consumer market. The essential elements are in place for the emergence and growth of an innovation led industry cluster in agriculture and food along the Colorado Front Range.

这些因素包括该地区人才库的质量、公共研究机构的实力、初级农业生产的绝对数量、农业生产与其他经济活动的接近程度以及当地消费市场的规模。科罗拉多州前沿区域农业和食品创新产业集群的出现和发展，已经具备必要的条件。

Public recognition of this innovation cluster has understandably been slow. Perhaps it is because agriculture is still seen by many in Colorado as an established, mature, primary industry, largely confined to rural regions. Many see agriculture as tied to a more traditional way of life and engaged in the low margins business of producing a handful of well-known commodities. At the same time, popular images of innovation are often more closely associated with "hi-tech" industries, like semiconductors, software, or aerospace. There are other reasons that innovation in agriculture may get overlooked. Some of the most interesting innovation going on within the agricultural value chain are readily associated with other industries, such as biotechnology, clean tech, manufacturing, retail, or tourism. It may also be due in part to the size and complexity of the value chain, with interrelated industry sectors and a wide range of different production methods and technologies being utilized. It is difficult for the public—and even for many agricultural professionals—to recognize and account for all of the various businesses that work together to feed, clothe, and fuel today's consumers.

可以理解,公众对这一创新集群的认知度一直很低。也许是因为在科罗拉多州,农业仍然被许多人视为已建成的成熟的第一产业,主要还是在农村地区。许多人认为,农业与一种更为传统的生活方式紧密相连,从事着生产少量知名商品的低利润率业务。与此同时,创新的流行形象往往与半导体、软件或航空航天等"高科技"行业联系更为紧密。农业创新可能被忽视还有其他原因。在农业价值链中进行的一些最有趣的创新很容易与其他行业相关联,如生物技术、清洁技术、制造业、零售业或旅游业。这也可能部分是因为价值链的规模和复杂性,有相互关联的各工业部门还有各种不同的生产方法和技术的使用。对公众来说,甚至对许多农业专业人士来说,要认识到并解释那些共同为今天的消费者提供食物、衣服和燃料的各种各样的企业是很困难的。

Innovation in the agriculture and food value chain is of crucial importance. It is what arguably kept agricultural production ahead of global population growth during the 20th century, averting massive famine. Further innovation will be essential to achieving an additional 70 percent of global agricultural production by 2050 to meet the needs of a largely middle class population of 9 billion, while sustaining the quality of our landscapes and water resources.

农业和食品价值链的创新至关重要。在20世纪,农业生产一直领先于全球人口增长,从而避免了大规模饥荒。进一步的创新对于到2050年实现全球农业生产的70%,以满足90亿中产阶级人口的需求至关重要,同时维持我们的景观和水资源的质量。

Many potential efficiency gains and quality improvements can be made to the current system. Yet, many potential innovations go beyond single, one-off, incremental improvements and will necessarily involve multiple players bound together in complex supplier-buyer or competitor relationships.

目前的系统可以获得许多潜在的效率提高和质量改进。然而，许多潜在的创新超越了单一的、一次性的、渐进式的改进，必然会涉及多个参与者，他们和复杂的供应商-买家或竞争对手捆绑在一起。

The value chain of agriculture has strong internal connections. Innovative changes introduced in one part of the value chain can impact a number of others, either directly or indirectly. For example, an innovative restaurant chain with a new business model can create new requirements for its wholesale food suppliers. That, in turn, can affect what is profitable for farmers to produce, which, in turn, can change the kinds of fertilizers or animal health supplements that they utilize and, therefore, purchase from farm and veterinary suppliers. Because of these deep interdependencies, vertical and horizontal coordination can help the introduction of an innovation to succeed. A lack of coordination can stymie or stall innovation. One of the fundamental tenants of cluster theory is that vertical and horizontal coordination are enhanced when all of those involved are co-located within a single geographic region.

农业价值链具有很强的内在联系。价值链某一部分引入的创新变化可以直接或间接地影响其他许多方面。例如，具有新商业模式的创新连锁餐厅可以为其食品批发供应商创造新的需求。这反过来又会影响农民的生产利润，改变他们所使用的肥料或动物保健品的种类，从而改变他们从农场和兽医供应商那里购买的产品。由于这些深层次的相互依赖，垂直和水平的协调可以帮助创新的引入获得成功。缺乏协调会阻碍创新。集群理论的一个基本观点是，当所有参与者都同时位于一个地理区域内时，垂直和水平的协调将得到增强。

Innovation can drive Colorado's economic development in a number of ways. The most significant economic benefits result when new products and services are introduced that make life better for consumers. Other impacts come from reducing the costs—or increasing the efficiencies—of providing already familiar products and services. Yet, the activities of undertaking innovation are, in themselves, primary drivers of economic development—whether it be conducting R&D, consulting for clients, launching new ventures, or providing specialized business, legal, and financial services. These activities require highly-skilled and well-paid knowledge workers. When adopted in the market, an innovation can make profits for the company that introduces it. While many of these economic development benefits are realized globally, or at least nationally, they can particularly benefit the region where the innovator is located and its innovation activities are conducted.

创新可以在许多方面推动科罗拉多州的经济发展。最显著的经济效益来自新产品和新服务的推出，使消费者的生活更美好。其他的影响来自降低成本——或者提高效率——提供已经熟悉的产品和服务。然而，从事创新活动本身就是经济发展的主要驱动力——无论是进行研发、为客户提供咨询、创办新企业，还是提供专门的业务、法律和金融服务皆是如此。这些活动需要高技能和高薪酬的知识工作者。当一项创新在市场上被采用时，它可以为引进它的公司带来利润。虽然这些经济发展效益中有许多是在全球实现的，或至少是在全国实现的，但它们对创新者所在的地区和开展创新活动的地区尤其有利。

(http：//innovation. colostate. edu/ and http：//www. news. colostate. edu/content/documents/ ColoradoInnovationReport2014final. pdf)

练习
Exercises

练习一　将原课文做视译练习

练习二　阅读理解

Passage A

Even plants can run a fever, especially when they're under attack by insects or disease. But unlike humans, plants can have their temperature taken from 3,000 feet away—straight up. A decade ago, adapting the infrared （红外线） scanning technology developed for military purposes and other satellites, physicist Stephen Paley came up with a quick way to take the temperature of crops to determine which ones are under stress. The goal was to let farmers precisely target pesticide spraying rather than rain poison on a whole field, which invariably includes plants that don't have pest （害虫） problems.

Even better, Paley's Remote Scanning Services Company could detect crop problems before they became visible to the eye. Mounted on a plane flown at 3,000 feet at night, an infrared scanner measured the heat emitted by crops. The data were transformed into a color-coded map showing where plants were running "fevers". Farmers could then spot-spray, using 50 to 70 percent less pesticide than they otherwise would.

The bad news is that Paley's company closed down in 1984, after only three years. Farmers resisted the new technology and long-term backers were hard to find. But with the renewed concern about pesticides on produce, and refinements in infrared scanning, Paley hopes to get back into operation. Agriculture experts have no doubt the technology works. "This technique can be used on 75 percent of agricultural land in the United States," says George Oerther of Texas A&M. Ray Jackson, who recently retired from the Department of Agriculture, thinks remote infrared crop scanning could be adopted by the end of the decade. But only if Paley finds the financial backing which he failed to obtain 10 years ago.

1. Plants will emit an increased amount of heat when they are _____.
   A. sprayed with pesticides
   B. facing an infrared scanner
   C. in poor physical condition
   D. exposed to excessive sun rays

2. In order to apply pesticide spraying precisely, we can use infrared scanning to _____.

A. estimate the damage to the crops

B. draw a color-coded map

C. measure the size of the affected area

D. locate the problem area

3. Farmers can save a considerable amount of pesticide by _____.

A. resorting to spot-spraying

B. consulting infrared scanning experts

C. transforming poisoned rain

D. detecting crop problems at an early date

4. The application of infrared scanning technology to agriculture met with some difficulties due to _____.

A. the lack of official support

B. its high cost

C. the lack of financial support

D. its failure to help increase production

5. Infrared scanning technology may be brought back into operation because of _____.

A. the desire of farmers to improve the quality of their produce

B. growing concern about the excessive use of pesticides on crops

C. the forceful promotion by the Department of Agriculture

D. full support from agricultural experts

Passage B

The development of Jamestown in Virginia during the second half of the seventeenth century was closely related to the making and the use of bricks. There are several practical reasons why bricks become important to the colony. Although the forests could initially supply sufficient timber, the process of lumbering （采伐） was extremely difficult, particularly because of the lack of roads. Later, when the timber on the peninsula （半岛） had been depleted （耗尽）, wood had to be brought from some distance. Building stone was also in short supply. However, as clay was plentiful, it was inevitable that the colonists should turn to brick-making.

In addition to practical reason for using brick as the principal construction material, there was also an ideological reason. Brick represented durability and permanence. The Virginia Company of London instructed the colonists to build hospitals and new residences out of brick. In 1662, the Town Act of the Virginia Assembly provided for the construction of thirty-two brick buildings and prohibited the use of wood as a construction material. Had this law ever been successfully enforced, Jamestown would have been a model city. Instead, the residents failed to comply fully with the law, and by 1699 Jamestown had collapsed into a pile of rubble with only three or four habitable houses.

1. In the first half of 1600's most building in Jamestown were probably made of _____.
   A. earth   B. stone   C. wood   D. brick

2. Which of the following was NOT a reason for using brick in construction?
   A. Wood had to be brought from some distance.
   B. There was considerable clay available.
   C. The lumbering process depended on good roads.
   D. The timber was not of good quality.

3. It can be inferred from the passage that settlers who built with bricks in the 1600's were _____.
   A. planning to return to England
   B. not concerned about durability
   C. obeying the laws
   D. interested in large residences

4. It can be inferred from the passage that prior to the action of the Virginia Company of London, Jamestown had an insufficient number of _____.
   A. colonists   B. material facilities
   C. clay sources   D. bricklayers

5. According to the passage, what eventually happened to Jamestown?

A. It was practically destroyed.
B. It remained the seat of government.
C. It became a model city.
D. It was almost completed.

第二单元 安徽农业大学简介
UNIT TWO
A BRIEF INTRODUCTION TO
ANHUI AGRICULTURAL UNIVERSITY

安徽农业大学坐落于安徽省省会合肥,是一所办学历史悠久、学科门类齐全、社会影响广泛的省属重点高校。2011年成为安徽省人民政府和农业部共建高校。2012年成为国家首批建设"新农村发展研究院"的十所高校之一。2013年入选安徽省首批"有特色高水平"大学建设高校。2014年成为安徽省人民政府和国家林业局共建高校。

学校源于1928年成立的省立安徽大学,1935年成立农学院. 1954年独立办学,1995年更名为安徽农业大学。校园占地面积3528亩,建筑面积78万平方米,教学科研仪器设备总价值3.8亿元,图书馆藏书260万册。

学校下设18个学院。现有全日制普通在校生21177人,其中硕士、博士研究生2597人。现有在职教职工1680人,其中教授、副教授等高级专业技术职务610人,拥有博士学位383人,博士生、硕士生导师400多人。现有"国家杰青"1人,省"百人计划"2人,"皖江学者"3人,校聘特聘(讲席)教授30人。

学校有77个本科专业,其中5个国家级特色专业;2个国家级教学团队,2门国家级精品课程、1门国家级视频公开课程。现有1个国家重点(培育)学科和17个省部级重点学科,5个博士后科研流动站,5个一级学科博士学位授权点,32个二级学科博士学位授权点,18个一级学科硕士学位授权点,71个二级学科硕士学位授权点,7类专业学位硕士授权点。

学校现有1个国家重点实验室、33个省部级重点实验室、工程技术研究中心等省部级科研平台,2个省级"2011协同创新中心",1个教育部"长江学者和创新团队发展计划"创新团队,9个省级创新团队,6位国家现代农业产业技术体系岗位科学家。近五年来,学校主持国家级科研、教研项目333项,荣获省部级以上奖励88项,其中国家科技进步二等奖和省重大科技成就奖各1项，省科技进步一等奖4项，国家教学成果二等奖3项。

学校坚持开放办学，大力实施国际化办学战略，先后与美国、英国、加拿大、日本、德国、澳大利亚等30多个国家和地区的高校、科研院所建立了稳定的合作关系。2011年，学校获批为中国科协"安徽海智农业基地"和教育部"接受中国政府奖学金来华留学生院校"；2012年成为农业部"南南合作"项目外语培训定点单位；2014年获批为"国家示范型国际科技合作基地"。

学校始终坚持服务"三农"的办学方向，致力于科教兴农、科教兴皖事业，走出了一条享誉全国的富民、兴校、创新、育人的"大别山道路"，为推动现代农业发展和地方经济社会建设作出了突出贡献，得到党和国家领导人的充分肯定。

当前，学校正围绕建设特色鲜明的高水平农业大学的奋斗目标，全面深化改革，大力推进内涵发展、特色发展、创新发展、和谐发展，努力为服务美好安徽建设作出新的更大贡献。

Located in Hefei, capital city of Anhui Province, China, Anhui Agricultural University (AAU) is a key university with a long history, comprehensive disciplines and extensive social influences. It is supported by Anhui Provincial Government, jointly with China Ministry of Agriculture since 2011 and with China Bureau of Forestry since 2014. In 2012 it was approved by Chinese central government as one of the first ten universities nationwide to establish a Research Institute for New Rural Development, and in 2013 selected as one of the "distinctive universities with unique features" in Anhui Province.

AAU started its history from College of Agriculture in 1935 under Provincial Anhui University founded in 1928, it became an independent higher learning institution in 1954 and changed its name to Anhui Agricultural University since 1995. Its campus covers an area of 235 hectares, with a floor space of 780 thousand square meters. Its teaching facilities and research instruments value 380 million RMB and its library has a holding of 2.6 million book volumes.

AAU now has 18 schools, with a full-time enrollment of 21,177 students, 2,597 out of whom are postgraduates. Among the 1,680 teaching staffs, 610 have professorship or associate professorship, 400 out of whom work as PhD and master students' supervisors. One professor is awarded the title of "China Outstanding Youth Foundation Winner", two "100 Anhui Distinctive Scholars" and three "Wanjiang Scholars". AAU also has 30 specially appointed chair professors and guest chair professors

At present, there are 77 undergraduate programs, including 5 national-level ones of distinctive features, plus 2 national instructional teams, 2 national boutique courses, 1 national massive online open course. In terms of postgraduate education, there is 1 national key discipline candidate, 17 provincial key disciplines, 5 post-doctoral mobile research stations, 5 A-level and 32 B-level PhD programs, as well as 18 A-level, 71 B-level and 7 special master programs.

AAU has a very strong strength in research, possessing 1 national and 33 provincial key laboratories, 2 provincial "2011 synergic innovative centers", 1 innovation team under the "Yangtze River Scholar and Innovative Team Development Plan" supported by Ministry of Education of the PRC and 9 provincial-level teams. 6 scientists are appointed on the posts of national modern agricultural extension technology networks. During the recent five years, AAU has been granted 333 projects funded by Chinese central government and won 88 awards of provincial or ministerial levels, including 1 second prize of National Sci-tech Progress, 4 first prizes of Provincial Sci-tech Progress, and 1 Provincial Sci-tech Big Achievement Award, plus 3 second prizes of National Instructional Achievements.

With an open-minded strategy for educational internationalization, AAU has set up close links with universities and research institutes from over 30 countries and regions, such as U.S., UK, Canada, Japan, Germany and Australia. In 2011 AAU was ratified respectively by China Association of Science and Technology as one of "Anhui Overseas Talents' Agricultural Bases" and by Ministry of Education of the PRC as one of Chinese universities to admit international students funded by China

Scholarship. In 2012 China Ministry of Agriculture conferred to AAU one of its foreign language training centers for the South-South Cooperation Project. And a nationwide "International Demonstrative Sci-tech Cooperation Base" was approved in 2014 as well.

Firmly aimed at serving "Agriculture, Countryside and Farmers", AAU has always dedicated itself to developing rural areas via sci-technology and developing Anhui via sci-education, and has successfully created a well-known Dabieshan Road, integrated of advancing the university with instructional innovations and overcoming poverty. The achievements AAU has made are so outstanding that China Central Government leaders speak highly of its significant contributions to Anhui agricultural modernization and local economic growth.

At present, moving towards an international standard agricultural university with its own distinctive features, AAU is undertaking more reforms in all-rounded aspects and advancing itself even further with intrinsic values and synergic innovations, to provide better service and make greater contributions to its community.

人才培养
Instructional Innovations

学校以本科教育为主体，逐步扩大研究生教育，积极开展留学生教育和国际合作办学，大力发展专业学位教育和继续教育。在本科教育上，实施了"现代青年农场主培养计划""创新试验班"等卓越计划以及"小学期制""主辅修制"等改革，建立了"体系开放、机制灵活、渠道互通、选择多样"的人才培养体制。在研究生教育上，突出创新能力提升的核心目标，大力推进招生、培养、评价和奖助制度改革，不断调整专业与类型结构，逐步推进跨学科培养和多单位联合培养。

Based on its undergraduate education, AAU is expanding its postgraduate and international students' education while conducting joint instructional programs with foreign colleges. Regarding undergraduate education, initiatives like "Modern Young Farmers Syllabus", "Pilot

Innovative Program" and other plans for excellent professionals, as well as summer and winter sessions, and the "Major plus Minor" curriculum designs, all reflect a flexible, open-minded, multi-channeled and multi-optional instructional system.

Meanwhile, postgraduate education, focusing on improving students' innovative capabilities, has taken new initiatives in student recruitment, instructional innovations and evaluations, as well as scholarship awarding and financial aids, and thus timely updated its programs and categories by motivating its students to take interdisciplinary learning and joint education among different programs.

学科建设
Disciplinary Development

学校坚持以农林、生命类学科为优势和特色，以提高学科对产业的支撑能力为目标，不断凝练学科方向，调整学科结构，促进学科交叉融合，构建了多学科相互支撑、协调发展的"九大学科群"体系。坚持学科、学位点、团队、平台四位一体，不断创新学科建设管理体制机制，重点打造学校公共服务平台和学科公共平台，不断完善各级平台开放共享机制。同时，依托二级学科，加强各类研究型功能实验室平台建设，实现"人人进团队、人人有平台"的目标，整体提升学科建设水平和科技创新能力。

Taking agriculture and forestry and life sciences as its unique characteristics while aimed at increasing its supportive strength for agro-industries, AAU keeps updating its disciplinary orientations and divisions to promote their crossover and integration, and has eventually established "9 Interdisciplinary Clusters", merging disciplines, programs, faculty and platforms into one network, for a frequent upgrading of its administrative mechanism. On one hand, enormous efforts have been invested in completing public service platforms of different levels. Relying on the B-level disciplines, on the other hand, research lab platforms are rapidly developed, with an overall goal that "everyone belongs to a certain research team and is involved in a certain research project", as a guarantee measure for the University's further disciplinary development and scientific innovations.

国际交流
International Exchanges

学校牢固树立开放办学的理念，抢抓机遇、积极谋划，大力实施国际化战略，不断加大开放办学力度，形成了多形式、多层次、多渠道的国际交流与合作新格局，提升了学校的办学层次和办学水平，也为服务农业走出去战略作出了积极贡献。

With an open-minded education ideology and actively participated-in initiatives, AAU is striding forward rapidly on its way to internationalization and shaping a new pattern of multi-models, multi-layers and multi-channels in its foreign exchanges and collaborations. This globalization strategy not only evidently improves the University's own educational standard and quality, but meanwhile makes its unique contributions to the strategy of agricultural "going global".

生词和词组
New Words and Expressions

博士后科研流动站	post-doctoral mobile research stations
园艺学	horticulture
林业工程	forestry engineering
生物学	biology
作物学	crop science
畜牧学	animal husbandry
博士点	doctoral programs
一级学科博士点	A-level PhD programs
林学	forestry
生态学	ecology
二级学科博士点	B-level PhD programs
作物栽培学与耕作学	crop cultivation and farming

作物遗传育种	crop genetics and breeding
作物生理生态	crop physiology and ecology
作物生物技术	crop biotechnology
作物安全生产	crop production safety
作物信息学	crop informatics
区域农业发展	regional development in agriculture
茶学	tea science
果树学	pomology
蔬菜学	olericulture
森林保护学	forest protection
森林培育	silviculture
林木遗传育种	forestry genetics and breeding
森林经理学	forestry management
园林植物与观赏园艺	ornamental plants and horticulture
野生动植物保护与利用	wildlife protection and utilization
水土保持与荒漠化防治	water-soil conservation and desertification control
植物学	botany
动物学	zoology
生理学	physiology
水生生物学	hydrobiology
微生物学	microbiology
神经生物学	neurobiology
遗传学	genetics
发育生物学	developmental biology
细胞生物学	cell biology
生物化学与分子生物学	biochemistry and molecular biology
生物物理学	biophysics
农业昆虫与害虫防治	agricultural entomology and pest control
动物遗传育种与繁殖	animal genetic breeding and reproduction
木材科学与技术	wood science and technology
硕士点	master degree programs
一级学科硕士点	A-level master degree programs

环境科学与工程	environmental science and engineering
畜牧学	animal husbandry
兽医学	veterinary medicine
农林经济管理	agro-forestry economics and management
机械工程	mechanical engineering
农业工程	agricultural engineering
大气科学	atmospheric sciences
食品科学与工程	food science and engineering
工商管理	business administration
风景园林学	landscape architecture
草学	practaculture science
二级学科硕士点	B-level master degree programs
木材科学与技术	wood science and technology
植物病理学	plant pathology
特种经济动物饲养	special economic animal breeding
气象学	meteorology
专业学位硕士授权点	special master degree programs
农业推广	agricultural extension
金融	finance
国家级重点学科	candidate for national key disciplines
省部级重点学科	provincial/ministerial key disciplines

练习
Exercises

练习一　翻译课文（笔译、口译、视译）

练习二　将下列中文译成英文

一、博士后科研流动站
园艺学、林业工程、生物学、作物学、畜牧学

二、博士点
- 一级学科博士点

作物学、园艺学、林学、生物学、生态学

- 二级学科博士点

作物栽培学与耕作学、作物遗传育种、作物生理生态、作物生物技术、作物安全生产、作物信息学、区域农业发展、茶学、果树学、蔬菜学、森林保护学、森林培育、林木遗传育种、森林经理学、园林植物与观赏园艺、野生动植物保护与利用、水土保持与荒漠化防治、植物学、动物学、生理学、水生生物学、微生物学、神经生物学、遗传学、发育生物学、细胞生物学、生物化学与分子生物学、生物物理学、农业昆虫与害虫防治、动物遗传育种与繁殖、木材科学与技术、生态学

三、硕士点
- 一级学科硕士点

生物学、环境科学与工程、作物学、园艺学、农业资源与环境、植物保护、畜牧学、兽医学、林学、农林经济管理、机械工程、农业工程、大气科学、食品科学与工程、工商管理、生态学、风景园林学、草学

- 二级学科硕士点

茶学、木材科学与技术、作物遗传育种、作物栽培学与耕作学、植物病理学、农业昆虫与害虫防治、特种经济动物饲养、森林保护学、气象学、农业经济管理、动物遗传育种与繁殖、动物营养与饲料科学、果树学、土壤学、临床兽医学、园林植物与观赏园艺、马克思主义基本原理、思想政治教育、植物学、动物学、生理学、水生生物学、神经生物学、发育生物学、细胞生物学、生物化学与分子生物学、预防兽医学、生物物理学、微生物学、农业机械化工程、农药学、环境工程、遗传学、森林培育、基础兽医学、技术经济及管理、植物营养学、车辆工程、计算机应用技术、食品科学、水土保持与荒漠化防治、草业科学、林木遗传育种、森林经理学、野生动植物保护与利用、营养与食品卫生学、林业经济管理、土地资源管理、产业经济学、生态学、作物生理生态、作物生物技术、作物安全生产、机械设计及理论、环境科学、农产品加工及贮藏工程、作物信息学、蔬菜学、区域农业发展

- 专业学位硕士授权点

农业推广、兽医、林业、林业工程、食品工程、风景园林、金融

四、国家级重点学科（培育）

茶学

五、省部级重点学科

茶学、畜牧学、动物遗传育种与繁殖、木材科学与技术、森林保护学、生物物理学、植物病理学/农业昆虫与害虫防治、作物遗传育种、农业机械化工程、农业经济管理、遗传学、农产品加工及贮藏工程、作物栽培学与耕作学、果树学、农药学、预防兽医学、森林培育学

译文参考

Translation for Reference

  A. Post-Doctoral Mobile Research Stations

  Horticulture, Forestry Engineering, Biology, Crop Science, Animal Husbandry

  B. Doctoral Programs

  • A-Level PhD Programs

  Crop Science, Horticulture, Forestry, Biology, Ecology

  • B-Level PhD Programs

  Crop Cultivation and Farming, Crop Genetics and Breeding, Crop Physiology and Ecology, Crop Biotechnology, Crop Production Safety, Crop Informatics, Regional Development in Agriculture, Tea Science, Pomology, Olericulture, Forest Protection, Silviculture, Forestry Genetics and Breeding, Forestry Management, Ornamental Plants and Horticulture, Wildlife Protection and Utilization, Water-soil Conservation and Desertification Control, Botany, Zoology, Physiology, Hydrobiology, Microbiology, Neurobiology, Genetics Developmental Biology, Cell Biology, Biochemistry and Molecular Biology, Biophysics, Agricultural Entomology and Pest Control, Animal Genetic Breeding and Reproduction, Wood Science and Technology, Ecology

  C. Master Degree Programs

  • A-Level Master Degree Programs

  Biology, Environmental Science and Engineering, Crop Science,

Horticulture, Agricultural Resources and Utilization, Plant Protection, Animal Husbandry, Veterinary Medicine, Forestry, Agro-forestry Economics and Management, Mechanical Engineering, Agricultural Engineering, Atmospheric Sciences, Food Science and Engineering, Business Administration, Ecology, Landscape Architecture, Practaculture Science

• B-Level Master Degree Programs

Tea Science, Wood Science and Technology, Crop Genetics and Breeding, Crop Cultivation and Farming, Plant Pathology, Agricultural Entomology and Pest Control, Special Economic Animal Breeding, Forestry Protection, Meteorology, Agricultural Economics and Management, Animal Genetics, Breeding and Reproduction, Animal Nutrition and Feed Science, Pomology, Soil Science, Veterinary Medicine, Clinical Veterinary Science, Preventive Veterinary Medicine, Biophysics, Microbiology, Agricultural Mechanization and Engineering, Pesticide Science, Environmental Engineering, Genetics Silviculture, Technology Economics and Management, Plant Nutrition, Industrial Economics, Ecology, Mechanical Design and Theory, Environmental Science, AG Outputs Processing and Storage, Olericulture Ornamental Plants and Horticulture, Marxist Principles, Political Science, Botany, Zoology, Physiology, Hydrobiology, Neurobiology, Developmental Biology, Cell Biology, Biochemistry and Molecular Biology, Vehicle Engineering, Computer Technology, Food Science, Practaculture Science, Forestry Genetics and Breeding, Forestry Management, Wildlife Conservation and Utilization, Water-soil Conservation and Desertification Control Nutrition and Food Hygiene, Forestry Economics and Management, Land Resources and Management, Crop Physiology and Ecology, Crop Biotechnology, Crop Production Safety, Crop Information, Regional Development in Agriculture

• Special Master Degree Programs

Agricultural Extension, Veterinary Medicine, Forestry, Forestry Engineering, Food Engineering, Landscape Architecture, Finance

PART THREE INTERNATIONAL EXCHANGE

D. Candidate for National Key Disciplines

Tea Science

E. Provincial/Ministerial Key Disciplines

Tea Science, Animal Husbandry, Animal Genetics, Breeding and Reproduction, Wood Science and Technology, Forestry Protection, Biophysics, Plant Pathology/Agricultural Entomology and Pest Control, Crop Genetics and Breeding, Agricultural Mechanization and Engineering, Agricultural Economics and Management, Genetics, AG Outputs Processing and Storage, Crop Cultivation and Farming, Pomology, Pesticide Science, Preventive Veterinary Medicine, Silviculture

第三单元　共谋绿色生活，共建美丽家园

——在 2019 年中国北京世界园艺博览会开幕式上的讲话

2019 年 4 月 28 日，北京

习近平

UNIT THREE

WORKING TOGETHER FOR A GREEN AND BETTER FUTURE FOR ALL

——Remarks at the Opening Ceremony of the International Horticultural Exhibition 2019 Beijing China

Beijing, 28 April 2019

Xi Jinping

尊敬的各位国家元首，政府首脑和夫人，
尊敬的国际展览局秘书长和国际园艺生产者协会主席，
尊敬的各国使节，各位国际组织代表，
女士们，先生们，朋友们：

Your Excellencies Heads of State and Government and Your Spouses,
Your Excellency Secretary General of the Bureau International des Expositions,
Your Excellency President of the International Association of Horticultural Producers,
Your Excellencies Diplomatic Envoys and Representatives of International Organizations,
Ladies and Gentlemen,
Friends,

"迟日江山丽，春风花草香。"四月的北京，春回大地，万物复苏。很高兴同各位嘉宾相聚在雄伟的长城脚下、美丽的妫水河畔，共同拉开 2019 年中国北京世界园艺博览会大幕。

"Spring scenery greets the eye; sweet blooms perfume the air." Such is the delight of April in Beijing, full of exuberance and vitality, as depicted by an ancient Chinese poem. By the beautiful Guishui River and at the foot of the majestic Great Wall, we are very glad to welcome all the distinguished guests to the opening of the International Horticultural Exhibition 2019 Beijing China.

首先，我谨代表中国政府和中国人民，并以我个人的名义，对远道而来的各位嘉宾，表示热烈的欢迎！对支持和参与北京世界园艺博览会的各国朋友，表示衷心的感谢！

On behalf of the Chinese government and people and in my own name, I wish to extend a warm welcome to all the guests coming to the Expo and sincere appreciation to all the friends for their support and participation.

北京世界园艺博览会以"绿色生活，美丽家园"为主题，旨在倡导人们尊重自然、融入自然、追求美好生活。北京世界园艺博览会园区，同大自然的湖光山色交相辉映。我希望，这片园区所阐释的绿色发展理念能传导至世界各个角落。

The Expo, as indicated by its theme "Live Green, Live Better", aims to promote respect for nature and a better life in harmony with nature. It is our hope that this Expo Park, designed to blend into its splendid surrounding landscape, will show the world the vision of green development.

女士们、先生们、朋友们！

Ladies and Gentlemen,
Friends,

锦绣中华大地，是中华民族赖以生存和发展的家园，孕育了中华民族5000多年的灿烂文明，造就了中华民族天人合一的崇高追求。

This beautiful land of China, home to the Chinese nation and its splendid 5,000-year civilization, has nurtured the lofty idea of harmony between man and nature.

现在，生态文明建设已经纳入中国国家发展总体布局，建设美丽中国已经成为中国人民心向往之的奋斗目标。中国生态文明建设进入了快车道，天更蓝、山更绿、水更清将不断展现在世人面前。

Ecological conservation has become part of China's overall plan for national development. Building a beautiful China is an inspiring goal for the Chinese people. As China steps up its conservation efforts, the world will see a China with more blue skies, lush mountains and lucid waters.

纵观人类文明发展史，生态兴则文明兴，生态衰则文明衰。工业化进程创造了前所未有的物质财富，也产生了难以弥补的生态创伤。杀鸡取卵、竭泽而渔的发展方式走到了尽头，顺应自然、保护生态的绿色发展昭示着未来。

The history of civilizations shows that the rise or fall of a civilization is closely tied to its relationship with nature. Industrialization, while generating unprecedented material wealth, has incurred serious damage to Mother Nature. Development without thought to the future is not sustainable. The way forward should be green development that focuses on harmony with nature and eco-friendly progress.

女士们、先生们、朋友们！

Ladies and Gentlemen,

Friends,

仰望夜空，繁星闪烁。地球是全人类赖以生存的唯一家园。我们要像保护自己的眼睛一样保护生态环境，像对待生命一样对待生态环境，同筑生态文明之基，同走绿色发展之路！

Looking up at night, we are awed by the many stars in the sky. Yet, planet Earth is the only home for mankind. We must protect this planet like our own eyes, and cherish nature the way we cherish life. We must preserve what gives our planet life and embrace green development.

——我们应该追求人与自然和谐。山峦层林尽染，平原蓝绿交融，城乡鸟语花香。这样的自然美景，既带给人们美的享受，也是人类走向未来的依托。无序开发、粗暴掠夺，人类定会遭到大自然的无情报复；合理利用、友好保护，人类必将获得大自然的慷慨回报。我们要维持地球生态整体平衡，让子孙后代既能享有丰富的物质财富，又能遥望星空、看见青山、闻到花香。

——We need to advocate harmony between man and nature. Lush mountains, green fields, singing birds and blossoming flowers offer more than beauty to the eye. They are the basis for future development. Nature will punish those who exploit and plunder it brutally, and reward those who use and protect it carefully. We must maintain the overall balance of the Earth's eco-system, so that our children and children's children will not only have material wealth but also enjoy starry skies, green mountains and sweet flowers.

——我们应该追求绿色发展繁荣。绿色是大自然的底色。我一直讲，绿水青山就是金山银山，改善生态环境就是发展生产力。良好生态本身蕴含着无穷的经济价值，能够源源不断创造综合效益，实现经济社会可持续发展。

——We need to pursue prosperity through green development. Green is the color of nature. I have always said that green mountains and lucid waters are indeed mountains of gold and silver, and that environmental improvement means greater productivity. A sound environment promises great economic potential, generates good returns, and contributes to economic and social sustainability.

——我们应该追求热爱自然情怀。"取之有度，用之有节"，是生态文明的真谛。我们要倡导简约适度、绿色低碳的生活方式，拒绝奢华和浪费，形成文明健康的生活风尚。要倡导环保意识、生态意识，构建全社会共同参与的环境治理体系，让生态环保思想成为社会生活中的主流文化。要倡导尊重自然、爱护自然的绿色价值观念，让天蓝地绿水清深入人心，形成深刻的人文情怀。

——We need to follow a philosophy that cares for nature. Well-measured use of natural resources is the key to ecological conservation. We need to promote a simpler, greener and low-carbon lifestyle, oppose excessiveness and foster a culture of living green and living healthy. We need to raise people's awareness, develop a conservation system in which everyone plays a part, and make mainstream ecological conservation merge into every aspect of social life. We need to advocate the value of green development that reveres and cares for nature so that blue skies, green fields and clear waters will be a vision cherished by all.

——我们应该追求科学治理精神。生态治理必须遵循规律,科学规划,因地制宜,统筹兼顾,打造多元共生的生态系统。只有赋之以人类智慧,地球家园才会充满生机活力。生态治理,道阻且长,行则将至。我们既要有只争朝夕的精神,更要有持之以恒的坚守。

——We need to adopt a scientific approach to ecological conservation. To create an eco-system where all elements coexist in harmony, we need to follow the laws of nature, base our efforts on scientific planning, adopt a holistic approach to conservation, and factor in local conditions. Human wisdom is essential to sustaining the dynamism of Earth, our common homeland. Ecological conservation may be a long and arduous effort. We must press ahead with a sense of urgency and perseverance to achieve our goals.

——我们应该追求携手合作应对。建设美丽家园是人类的共同梦想。面对生态环境挑战,人类是一荣俱荣、一损俱损的命运共同体,没有哪个国家能独善其身。唯有携手合作,我们才能有效应对气候变化、海洋污染、生物保护等全球性环境问题,实现联合国2030年可持续发展目标。只有并肩同行,才能让绿色发展理念深入人心、全球生态文明之路行稳致远。

——We need to join hands to meet common challenges. A beautiful homeland is the shared aspiration of mankind. In the face of environmental challenges, all countries are in a community with destinies linked, and no country can stay immune. Only together can we effectively address climate change, marine pollution, biological conservation and other global environmental issues and achieve the UN 2030 Agenda for Sustainable

Development. Only concerted efforts can drive home the idea of green development and bring about steady progress in the ecological conservation of the globe.

女士们、先生们、朋友们！

Ladies and Gentlemen，Friends，

  昨天，第二届"一带一路"国际合作高峰论坛成功闭幕，在座许多嘉宾出席了论坛。共建"一带一路"就是要建设一条开放发展之路，同时也必须是一条绿色发展之路。这是与会各方达成的重要共识。中国愿同各国一道，共同建设美丽地球家园，共同构建人类命运共同体。

  Many of you attended the Second Belt and Road Forum for International Cooperation, which drew to a successful conclusion yesterday. It was agreed at the Forum that we will pursue open and green development in Belt and Road cooperation. China is ready to work with all other countries to build a better homeland and a community with a shared future for mankind.

女士们、先生们、朋友们！

Ladies and Gentlemen，Friends，

  一代人有一代人的使命。建设生态文明，功在当代，利在千秋。让我们从自己、从现在做起，把接力棒一棒一棒传下去。

  Every generation has its own mission. Our efforts to conserve the ecosystem will benefit not only this generation, but many more to come. Let us act now, start with ourselves, and make sure that the baton of conservation will be passed on.

  我宣布，2019年中国北京世界园艺博览会开幕！

  I now declare open the International Horticultural Exhibition 2019 Beijing China!

生词和词组
New Words and Expressions

2019年中国北京世界园艺博览会	International Horticultural Exhibition 2019 Beijing China
妫水	Guishui（River）
交相辉映	add radiance and color to each other
杀鸡取卵	kill the goose that lays the golden eggs
竭泽而渔	drain the pond to get all the fish
鸟语花香	singing birds and blossoming flowers
绿水青山	green mountains and lucid waters
独善其身	stay immune

翻译探讨
Translation Study and Discussion

  中文：共谋绿色生活，共建美丽家园

  英文：Working Together for a Green and Better Future for All

  提示：中文标题往往是对仗、对偶句形式，英文则采取意译方式。注意英文标题格式的书写方式，简而言之，首词和实词需大写，整个标题一般为名词化形式。

  中文：尊敬的……

  英文：英文尊称或敬语"Your Excellency"，"Your Excellencies"

  提示：除了 respectable, distinguished 表示"尊敬的"，还有"Your Excellency"，"Your Excellencies"表示尊称或敬语，具体单复数由其后人员数确定。

  中文："迟日江山丽，春风花草香。"四月的北京，春回大地，万物复苏。

  英文："Spring scenery greets the eye; sweet blooms perfume the air." Such is the delight of April in Beijing, full of exuberance and vitality, as depicted by an ancient Chinese poem.

第三部分 国际交流
PART THREE INTERNATIONAL EXCHANGE

提示：演讲稿中往往涉及诗词的翻译，应"音""形""意"之美尽显，"信""达""雅"具备。如此句，"spring"和"sweet"为头韵，两句保留了原诗句的并列结构，英文皆为主谓宾短句型；同时该句还进行了增译，考虑到受众，增加了解释性词语"as depicted by an ancient Chinese poem"说明。

中文：首先，我谨代表中国政府和中国人民，并以我个人的名义，对远道而来的各位嘉宾，表示热烈的欢迎！对支持和参与北京世界园艺博览会的各国朋友，表示衷心的感谢！

英文：On behalf of the Chinese government and people and in my own name, I wish to extend a warm welcome to all the guests coming to the Expo and sincere appreciation to all the friends for their support and participation.

提示：原文两个句子用两次"表示"，英文用一次"extend"，为合译。

中文：北京世界园艺博览会以"绿色生活，美丽家园"为主题，旨在倡导人们尊重自然、融入自然、追求美好生活。

英文：The Expo, as indicated by its theme "Live Green, Live Better", aims to promote respect for nature and a better life in harmony with nature.

提示：主题原中文为名词性词组，英译为祈使句，名词转译为动词。

中文：锦绣中华大地，是中华民族赖以生存和发展的家园，孕育了中华民族5000多年的灿烂文明，造就了中华民族天人合一的崇高追求。

英文：This beautiful land of China, home to the Chinese nation and its splendid 5,000-year civilization, has nurtured the lofty idea of harmony between man and nature.

提示：这里，"是"字句译为相关同位语结构，作为主语的一部分。

中文：中国生态文明建设进入了快车道，天更蓝、山更绿、水更清将不断展现在世人面前。

英文：As China steps up its conservation efforts, the world will see a China with more blue skies, lush mountains and lucid waters.

提示：注意比较级在语境中的应用。

中文：纵观人类文明发展史，生态兴则文明兴，生态衰则文明衰。

英文：The history of civilizations shows that the rise or fall of a civilization is closely tied to its relationship with nature.

提示：几个分句英译后为一个句子，为合译。

中文：工业化进程创造了前所未有的物质财富，也产生了难以弥补的生态创伤。

英文：Industrialization, while generating unprecedented material wealth, has incurred serious damage to Mother Nature.

提示：汉语意合而英语形合。分析原文上下文关系为"虽然……但是"，英文作相应处理。

中文：杀鸡取卵、竭泽而渔的发展方式走到了尽头，顺应自然、保护生态的绿色发展昭示着未来。

英文：Development without thought to the future is not sustainable. The way forward should be green development that focuses on harmony with nature and eco-friendly progress.

提示：此处成语翻译为意译，"舍形取义"。

中文：无序开发、粗暴掠夺，人类定会遭到大自然的无情报复；合理利用、友好保护，人类必将获得大自然的慷慨回报。

英文：Nature will punish those who exploit and plunder it brutally, and reward those who use and protect it carefully.

提示：注意英译主语的转换。汉语句中的主语是"人类"，英文则是"nature"。

中文：我们应该追求热爱自然情怀。"取之有度，用之有节"，是生态文明的真谛。

英文：We need to follow a philosophy that cares for nature. Well-measured use of natural resources is the key to ecological conservation.

提示："自然情怀"译为"philosophy"；"取之有度，用之有节"译为"Well-measured use of（natural resources）"。注意措辞的准确性和语言转换的灵活性。

中文：我们应该追求科学治理精神。
英文：We need to adopt a scientific approach to ecological conservation.

提示："精神"不都译为"spirit（s）"。所有的遣词造句都是根据上下文而做的选择。

中文：生态治理必须遵循规律，科学规划，因地制宜，统筹兼顾，打造多元共生的生态系统。
英文：To create an eco-system where all elements coexist in harmony, we need to follow the laws of nature, base our efforts on scientific planning, adopt a holistic approach to conservation, and factor in local conditions.

提示：注意英译中逻辑关系的还原、语序的变化、主语"we"的添加。

中文：我们既要有只争朝夕的精神，更要有持之以恒的坚守。
英文：We must press ahead with a sense of urgency and perseverance to achieve our goals.

提示：这里"精神"依然未译为"spirit"，但是意义已然蕴含在译文"a sense of urgency and perseverance"中。

中文：建设美丽家园是人类的共同梦想。
英文：A beautiful homeland is the shared aspiration of mankind.

提示："共同梦想"可以是"common dream"，但不拘泥于此，如这里的"shared aspiration"。

中文：共建"一带一路"就是要建设一条开放发展之路，同时也必须是一条绿色发展之路。

英文：It was agreed at the Forum that we will pursue open and green development in Belt and Road cooperation.

提示：此句为合译，包含"开放""绿色""发展"所有内涵。

中文：只有并肩同行，才能让绿色发展理念深入人心、全球生态文明之路行稳致远。

英文：Only concerted efforts can drive home the idea of green development and bring about steady progress in the ecological conservation of the globe.

提示：注意全文成语方面、四字格方面的英文翻译。这里仍然采取意译方式，"舍形取义"。"并肩同行"为"concerted efforts""让……行稳致远"为"bring about steady progress in…"。

练习
Exercises

练习一　翻译课文（笔译、口译、视译）

练习二　阅读理解

Passage A

A futures contract is simply a legally binding standardized agreement to deliver or take delivery of a given quality and quantity of a commodity at an agreed price at a specific date and place. One who goes "short" in the market is simply synonymous with one who sells a contract, agreeing to make a delivery in a designated month. Going "long" is synonymous with one who buys a contract, agreeing to accept a delivery in a designated month. Its initial purpose was to allow a buyer the chance to better plan his or her business by protecting against price fluctuations.

Buying cattle futures contracts is basically, therefore, a bet on the future value of the commodity. Although future's buying is sometimes used as a hedge by cattle ranchers, commodity buyers and sellers against future price fluctuations, it is a very risky business. There are estimates that 75% to 95% of individual investors lose money in commodity futures markets. However, many firms and other large corporate agribusiness and financial interests have gained an advantageous "insider" position from "secret signals". Big commercial feedlots, grain companies and meat packers—who have lower production costs than farmer-feeders or can shift costs from one level to another—would start selling their futures when the futures price exceeded their costs. Such costs, however, were still below the level of those costs, which most other farmer-feeders could afford to hedge against a loss. Immediately a select number of commodity brokers and officers joined the selling and thus the combined selling pressure became so great the cattle prices immediately began to drop.

The net result of this action by these "insiders" was that it placed an artificial cap on prices at the farmer-feeder, break-even level that in fact reflected neither supply or demand nor the prevailing cash market price for cattle. Instead of cattle futures at the time permitting farmer-feeders to shift price risks by hedging in futures, the result of insider manipulation is that the market puts them at an exaggerated disadvantage to the big commercial feedlots who could hedge at the lower cost level. Often, cash market cattle prices either stay the same or go up. This is strong evidence that futures prices are artificial and not reflecting supply-demand conditions.

1. Why do we have futures contracts?
 A. To take delivery of a given commodity.
 B. To ensure delivery of a given commodity at market prices.
 C. To allow for some price certainty before a commodity is delivered.
 D. To stabilize the market against price fluctuations.

2. The main reason that futures are risky in the cattle industry is due to _____.
 A. price fluctuations in the price of cattle

B. dropping prices of cattle

C. the uncertainty of the economy

D. insiders manipulating the prices

3. Why don't futures prices reflect the market price of cattle?

A. Because bigger companies with lower costs force down future prices.

B. Because futures were overpriced previously.

C. Because big companies ensure that futures were higher than the price of cattle.

D. Because of the oversupply of futures.

4. The author shows that cattle prices went down as the result of _____.

A. the mass sell-off of cattle futures

B. people selling their cattle to cover their losses

C. big commercial feedlots selling their cattle

D. a drop in demand for cattle

5. What advantage did the huge sell-off of futures give to big commercial feedlots?

A. It allowed them to charge high prices for cattle.

B. It allowed them to sell futures at artificially low rates.

C. It allowed them to push down the price of cattle.

D. It allowed them to buy out smaller farm feeders.

Passage B

Cuttlefish are intriguing little animals, and it's best not to have tunnel vision towards them, merely because they may first seem a bit monster-like. The cuttlefish resembles a rather large squid and is like the octopus, a member of the order of cephalopods. Although they are not considered the most highly evolved of the cephalopods, they are extremely intelligent.

On a scuba dive when you're observing cuttlefish, it is best to move slowly because cuttlefish have excellent eyesight and will probably see you first. In fact, their eyes may seem like the windows of your soul, for while observing them, it is hard to tell who is doing the observing, you or the cuttlefish. They are also highly mobile and fast creatures and come equipped with a small jet located just below the tentacles, which can expel water to help them move. Ribbons of flexible fin on each side of the body allow cuttlefish to hover, move, stop, and start.

By far their most intriguing characteristic is their ability to change body color and pattern. The cuttlefish is sometimes referred to as the "chameleon of the sea" because it can change its skin color and pattern instantaneously. Masters of camouflage, they can blend into any environment for protection, but they are also capable of the most imaginative displays of iridescent, brilliant color and intricate designs, which scientists believe they use to communicate with each other and for mating displays. However, judging from the riot of ornaments and hues cuttlefish produce, it is hard not to believe they paint themselves so beautifully just for the sheer joy of it. At the least, cuttlefish conversion must be the most sparkling in all the sea.

1. Which of the following is correct according to the information given in the text?
    A. Cuttlefish are a type of squid.
    B. Cuttlefish use jet propulsion as one form of locomotion.
    C. The Cuttlefish does not have an exoskeleton.
    D. Cuttlefish are the most intelligent cephalopods.
2. By saying "their eyes may seem like the windows of your soul" (Line 3, Para. 2) the author means _____.
    A. cuttlefish make you feel guilty about invading their territory
    B. you may learn something about yourself from them
    C. their eyes show a certain beauty about them
    D. their eyes resemble yours, forcing you to be self-conscious
3. The text mainly deals with _____.
    A. why cuttlefish are intriguing and the communication skills

of cuttlefish

B. classification and difficulties of observing cuttlefish and how cuttlefish communicate

C. the cuttlefish's method of locomotion and color displays in mating behavior

D. the characteristics of cuttlefish, particularly their ability to change color

4. The word "riot" (Line 8, Para. 3) most probably means _____.

A. mixture　　B. display　　C. disorder　　D. pattern

5. It may be inferred from the text that the author's purpose is _____.

A. to prove the intelligence of cuttlefish

B. to explain the communication habits of cuttlefish

C. to produce a fanciful description of the "Chameleon of the sea"

D. to describe the "chameleon of the sea" informatively and entertainingly

第三部分　国际交流

PART THREE　INTERNATIONAL EXCHANGE | 175

第四单元　齐心开创共建"一带一路"美好未来

——在第二届"一带一路"国际合作高峰论坛开幕式上的主旨演讲

2019年4月26日，北京

习近平

UNIT FOUR
WORKING TOGETHER TO DELIVER A BRIGHTER FUTURE FOR BELT AND ROAD COOPERATION

Keynote Speech at the Opening Ceremony of the Second Belt
and Road Forum for International Cooperation

Beijing, 26 April 2019

Xi Jinping

尊敬的各位国家元首，政府首脑，

Your Excellencies Heads of State and Government,

各位高级代表，

Your Excellencies High-level Representatives,

各位国际组织负责人，

Your Excellencies Heads of International Organizations,

女士们，先生们，朋友们：

Ladies and Gentlemen, Friends,

上午好！"春秋多佳日，登高赋新诗。"在这个春意盎然的美好时节，我很高兴同各位嘉宾一道，共同出席第二届"一带一路"国际合作高峰论坛。首先，我谨代表中国政府和中国人民，并以我个人的名义，对各位来宾表示热烈的欢迎！

Good morning! As a line of a classical Chinese poem goes, "Spring and autumn are lovely seasons in which friends get together to climb up mountains and write poems." On this beautiful spring day, it gives me great pleasure to have you with us here at the Second Belt and Road Forum for International Cooperation (BRF). On behalf of the Chinese government and people and in my own name, I extend a very warm welcome to you all!

两年前,我们在这里举行首届高峰论坛,规划政策沟通、设施联通、贸易畅通、资金融通、民心相通的合作蓝图。今天,来自世界各地的朋友再次聚首。我期待着同大家一起,登高望远,携手前行,共同开创共建"一带一路"的美好未来。

Two years ago, it was here that we met for the First Belt and Road Forum for International Cooperation, where we drew a blueprint of cooperation to enhance policy, infrastructure, trade, financial and people-to-people connectivity. Today, we are once again meeting here with you, friends from across the world. I look forward to scaling new heights with you and enhancing our partnership. Together, we will create an even brighter future for Belt and Road cooperation.

同事们、朋友们!
Dear Colleagues and Friends,

共建"一带一路"倡议,目的是聚焦互联互通,深化务实合作,携手应对人类面临的各种风险挑战,实现互利共赢、共同发展。在各方共同努力下,"六廊六路多国多港"的互联互通架构基本形成,一大批合作项目落地生根,首届高峰论坛的各项成果顺利落实,150多个国家和国际组织同中国签署共建"一带一路"合作协议。共建"一带一路"倡议同联合国、东盟、非盟、欧盟、欧亚经济联盟等国际和地区组织的发展和合作规划对接,同各国发展战略对接。

The joint pursuit of the Belt and Road Initiative (BRI) aims to enhance connectivity and practical cooperation. It is about jointly meeting various challenges and risks confronting mankind and delivering win-win outcomes and common development. Thanks to the joint efforts of all of us involved in this initiative, a general connectivity framework consisting of six corridors, six connectivity routes and multiple countries and ports has been put in place. A large number of cooperation projects have been launched, and the decisions of the first BRF have been smoothly implemented. More than 150 countries and international organizations have signed agreements on Belt and Road cooperation with China. The complementarity between the BRI and the development plans or cooperation initiatives of international and regional organizations such as the United Nations, the Association of Southeast Asian Nations, the African Union, the European Union, the Eurasian Economic Union and between the BRI and the development strategies of the participating countries has been enhanced.

从亚欧大陆到非洲、美洲、大洋洲，共建"一带一路"为世界经济增长开辟了新空间，为国际贸易和投资搭建了新平台，为完善全球经济治理拓展了新实践，为增进各国民生福祉作出了新贡献，成为共同的机遇之路、繁荣之路。事实证明，共建"一带一路"不仅为世界各国发展提供了新机遇，也为中国开放发展开辟了新天地。

From the Eurasian continent to Africa, the Americas and Oceania, Belt and Road cooperation has opened up new space for global economic growth, produced new platforms for international trade and investment and offered new ways for improving global economic governance. Indeed, this initiative has helped improve people's lives in countries involved and created more opportunities for common prosperity. What we have achieved amply demonstrates that Belt and Road cooperation has both generated new opportunities for the development of all participating countries and opened up new horizon for China's development and opening-up.

中国古人说："万物得其本者生，百事得其道者成。"共建"一带一路"，顺应经济全球化的历史潮流，顺应全球治理体系变革的时代要求，顺应各国人民过上更好日子的强烈愿望。面向未来，我们要聚焦重点、深耕细作，共同绘制精谨细腻的"工笔画"，推动共建"一带一路"沿着高质量发展方向不断前进。

An ancient Chinese philosopher observed that "plants with strong roots grow well, and efforts with the right focus will ensure success." The

Belt and Road cooperation embraces the historical trend of economic globalization, responds to the call for improving the global governance system and meets people's longing for a better life. Going ahead, we should focus on priorities and project execution, move forward with results-oriented implementation, just like an architect refining the blueprint, and jointly promote high-quality Belt and Road cooperation.

——我们要秉持共商共建共享原则，倡导多边主义，大家的事大家商量着办，推动各方各施所长、各尽所能，通过双边合作、三方合作、多边合作等各种形式，把大家的优势和潜能充分发挥出来，聚沙成塔、积水成渊。

We need to be guided by the principle of extensive consultation, joint contribution and shared benefits. We need to act in the spirit of multilateralism, pursue cooperation through consultation and keep all participants motivated. We may, by engaging in bilateral, trilateral and multilateral cooperation, fully tap into the strengths of all participants. Just as a Chinese proverb says, "A tower is built when soil on earth accumulates, and a river is formed when streams come together."

——我们要坚持开放、绿色、廉洁理念，不搞封闭排他的小圈子，把绿色作为底色，推动绿色基础设施建设、绿色投资、绿色金融，保护好我们赖以生存的共同家园，坚持一切合作都在阳光下运作，共同以零容忍态度打击腐败。我们发起了《廉洁丝绸之路北京倡议》，愿同各方共建风清气正的丝绸之路。

We need to pursue open, green and clean cooperation. The Belt and Road is not an exclusive club; it aims to promote green development. We may launch green infrastructure projects, make green investment and provide green financing to protect the Earth which we all call home. In pursuing Belt and Road cooperation, everything should be done in a transparent way, and we should have zero tolerance for corruption. The "Beijing Initiative for Clean Silk Road" has been launched, which represents our strong commitment to transparency and clean governance in pursuing Belt and Road cooperation.

——我们要努力实现高标准、惠民生、可持续目标，引入各方普遍支持的规则标准，推动企业在项目建设、运营、采购、招投标等环节按照普遍接受的国际规则标准进行，同时要尊重各国法律法规。要坚持以人民为中心的发展思想，聚焦消除贫困、增加就业、改善民生，让共建"一带一路"成果更好惠及全体人民，为当地经济社会发展作出实实在在的贡献，同时确保商业和财政上的可持续性，做到善始善终、善作善成。

We need to pursue high standard cooperation to improve people's lives and promote sustainable development. We will adopt widely accepted rules and standards and encourage participating companies to follow general international rules and standards in project development, operation, procurement and tendering and bidding. The laws and regulations of participating countries should also be respected. We need to take a people-centered approach, give priority to poverty alleviation and job creation to see that the joint pursuit of Belt and Road cooperation will deliver true benefits to the people of participating countries and contribute to their social and economic development. We also need to ensure the commercial and fiscal sustainability of all projects so that they will achieve the intended goals as planned.

同事们、朋友们！
Dear Colleagues and Friends,

共建"一带一路"，关键是互联互通。我们应该构建全球互联互通伙伴关系，实现共同发展繁荣。我相信，只要大家齐心协力、守望相助，即使相隔万水千山，也一定能够走出一条互利共赢的康庄大道。

Connectivity is vital to advancing Belt and Road cooperation. We need to promote a global partnership of connectivity to achieve common development and prosperity. I am confident that as we work closely together, we will transcend geographical distance and embark on a path of win-win cooperation.

基础设施是互联互通的基石，也是许多国家发展面临的瓶颈。建设高质量、可持续、抗风险、价格合理、包容可及的基础设施,有利于各国充分发挥资源禀赋,更好融入全球供应链、产业链、价值链,实现联动发展。中国将同各方继续努力,构建以新亚欧大陆桥等经济走廊为引领,以中欧班列、陆海新通道等大通道和信息高速路为骨架,以铁路、港口、管网等为依托的互联互通网络。

Infrastructure is the bedrock of connectivity, while the lack of infrastructure has held up the development of many countries. High-quality, sustainable, resilient, affordable, inclusive and accessible infrastructure projects can help countries fully leverage their resource endowment, better integrate into the global supply, industrial and value chains, and realize inter-connected development. To this end, China will continue to work with other parties to build a connectivity network centering on economic corridors such as the New Eurasian Land Bridge, supplemented by major transportation routes like the China-Europe Railway Express and the New International Land-Sea Trade Corridor and information expressway, and reinforced by major railway, port and pipeline projects.

我们将继续发挥共建"一带一路"专项贷款、丝路基金、各类专项投资基金的作用,发展丝路主题债券,支持多边开发融资合作中心有效运作。我们欢迎多边和各国金融机构参与共建"一带一路"投融资,鼓励开展第三方市场合作,通过多方参与实现共同受益的目标。

We will continue to make good use of the Belt and Road Special Lending Scheme, the Silk Road Fund, and various special investment funds, develop Silk Road theme bonds, and support the Multilateral Cooperation Center for Development Finance in its operation. We welcome the participation of multilateral and national financial institutions in BRI investment and financing and encourage third-market cooperation. With the involvement of multiple stakeholders, we can surely deliver benefits to all.

商品、资金、技术、人员流通,可以为经济增长提供强劲动力和广阔空间。"河海不择细流,故能就其深。"如果人为阻断江河的流入,再大的海,迟早都有干涸的一天。我们要促进贸易和投资自由化便利化,旗帜鲜明反对保护主义,推动经济全球化朝着更加开放、包容、普惠、平衡、共赢的方向发展。

The flow of goods, capital, technology and people will power economic growth and create broad space for it. As a Chinese saying goes, "The ceaseless inflow of rivers makes the ocean deep." However, were such inflow to be cut, the ocean, however big, would eventually dry up. We need to promote trade and investment liberalization and facilitation, say no to protectionism, and make economic globalization more open, inclusive, balanced and beneficial to all.

我们将同更多国家商签高标准自由贸易协定,加强海关、税收、审计监管等领域合作,建立共建"一带一路"税收征管合作机制,加快推广"经认证的经营者"国际互认合作。我们还制定了《"一带一路"融资指导原则》,发布了《"一带一路"债务可持续性分析框架》,为共建"一带一路"融资合作提供指南。中方今年将举办第二届中国国际进口博览会,为各方进入中国市场搭建更广阔平台。

To this end, we will enter into negotiation with more countries to conclude high-standard free trade agreements, and strengthen cooperation in customs, taxation and audit oversight by setting up the Belt and Road Initiative Tax Administration Cooperation Mechanism and accelerating international collaboration on the mutual recognition of Authorized Economic Operators. We have also formulated the "Guiding Principles on Financing the Development of the Belt and Road" and published the "Debt Sustainability Framework for Participating Countries of the Belt and Road Initiative" to provide guidance for BRI financing cooperation. In addition, the Second China International Import Expo will be held this year to build an even bigger platform for other parties to access the Chinese market.

创新就是生产力,企业赖之以强,国家赖之以盛。我们要顺应第四次工业革命发展趋势,共同把握数字化、网络化、智能化发展机遇,共同探索新技术、新业态、新模式,探寻新的增长动能和发展路径,建设数字丝绸之路、创新丝绸之路。

Innovation boosts productivity; it makes companies competitive and countries strong. We need to keep up with the trend of the Fourth Industrial Revolution, jointly seize opportunities created by digital, networked and smart development, explore new technologies and new forms and models of business, foster new growth drivers and explore new development pathways, and build the digital Silk Road and the Silk Road of innovation.

中国将继续实施共建"一带一路"科技创新行动计划,同各方一道推进科技人文交流、共建联合实验室、科技园区合作、技术转移四大举措。我们将积极实施创新人才交流项目,未来5年支持5000人次中外方创新人才开展交流、培训、合作研究。我们还将支持各国企业合作推进信息通信基础设施建设,提升网络互联互通水平。

China will continue to carry out the Belt and Road Science, Technology and Innovation Cooperation Action Plan, and will work with our partners to pursue four major initiatives, namely the Science and Technology People-to-People Exchange Initiative, the Joint Laboratory Initiative, the Science Park Cooperation Initiative, and the Technology Transfer Initiative. We will actively implement the Belt and Road Initiative Talents Exchange Program, and will, in the coming five years, offer 5,000 opportunities for exchange, training and cooperative research for talents from China and other BRI participating countries. We will also support companies of various countries in jointly advancing ICT infrastructure building to upgrade cyber connectivity.

发展不平衡是当今世界最大的不平衡。在共建"一带一路"过程中,要始终从发展的视角看问题,将可持续发展理念融入项目选择、实施、管理的方方面面。我们要致力于加强国际发展合作,为发展中国家营造更多发展机遇和空间,帮助他们摆脱贫困,实现可持续发展。为此,我们同各方共建"一带一路"可持续城市联盟、绿色发展国际联盟,制定《"一带一路"绿色投资原则》,发起"关爱儿童、共享发展,促进可持续发展目标实现"合作倡议。我们启动共建"一带一路"生态环保大数据服务平台，将继续实施绿色丝路使者计划，并同有关国家一道，实施"一带一路"应对气候变化南南合作计划。我们还将深化农业、卫生、减灾、水资源等领域合作，同联合国在发展领域加强合作，努力缩小发展差距。

Imbalance in development is the greatest imbalance confronting today's world. In the joint pursuit of the BRI, we must always take a development-oriented approach and see that the vision of sustainable development underpins project selection, implementation and management. We need to strengthen international development cooperation so as to create more opportunities for developing countries, help them eradicate poverty and achieve sustainable development. In this connection, China and its partners have set up the Belt and Road Sustainable Cities Alliance and the BRI International Green Development Coalition, formulated the "Green Investment Principles for the Belt and Road Development", and launched the Declaration on Accelerating the Sustainable Development Goals for Children through Shared Development. We have set up the BRI Environmental Big Data Platform. We will continue to implement the Green Silk Road Envoys Program and work with relevant countries to jointly implement the Belt and Road South-South Cooperation Initiative on Climate Change. We will also deepen cooperation in agriculture, health, disaster mitigation and water resources; and we will enhance development cooperation with the United Nations to narrow the gap in development.

我们要积极架设不同文明互学互鉴的桥梁，深入开展教育、科学、文化、体育、旅游、卫生、考古等各领域人文合作，加强议会、政党、民间组织往来，密切妇女、青年、残疾人等群体交流，形成多元互动的人文交流格局。未来5年，中国将邀请共建"一带一路"国家的政党、智库、民间组织等1万名代表来华交流。

We need to build bridges for exchanges and mutual learning among different cultures, deepen cooperation in education, science, culture, sports, tourism, health and archaeology, strengthen exchanges between parliaments, political parties and non-governmental organizations and exchanges between women, young people and people with disabilities in order to facilitate multi-faceted people-to-people exchanges. To this end, we will, in the coming five years, invite 10,000 representatives of political parties, think tanks and non-governmental organizations from Belt and Road participating countries to visit China.

我们将鼓励和支持沿线国家社会组织广泛开展民生合作，联合开展一系列环保、反腐败等领域培训项目，深化各领域人力资源开发合作。我们将持续实施"丝绸之路"中国政府奖学金项目，举办"一带一路"青年创意与遗产论坛、青年学生"汉语桥"夏令营等活动。我们还将设立共建"一带一路"国际智库合作委员会、新闻合作联盟等机制，汇聚各方智慧和力量。

We will encourage and support extensive cooperation on livelihood projects among social organizations of participating countries, conduct a number of environmental protection and anti-corruption training courses and deepen human resources development cooperation in various areas. We will continue to run the Chinese government scholarship Silk Road Program, host the International Youth Forum on Creativity and Heritage along the Silk Roads and the Chinese Bridge summer camps. We will also put in place new mechanisms such as the Belt and Road Studies Network and the Belt and Road News Alliance to draw inspiration and pool our strength for greater synergy.

同事们、朋友们！
Dear Colleagues and Friends,

今年是中华人民共和国成立70周年。70年前，中国人民历经几代人上下求索，终于在中国共产党领导下建立了新中国，中国人民从此站了起来，中国人民的命运从此掌握在了自己手中。

This year marks the 70th anniversary of the founding of the People's Republic of China. Seven decades ago, through the arduous struggle carried out by several generations of Chinese people and under the leadership of the Communist Party of China, New China was founded. We

Chinese have since stood up and held our future in our own hands.

历经70年艰苦奋斗，中国人民立足本国国情，在实践中不断探索前进方向，开辟了中国特色社会主义道路。今天的中国，已经站在新的历史起点上。我们深知，尽管成就辉煌，但前方还有一座座山峰需要翻越，还有一个个险滩等待跋涉。我们将继续沿着中国特色社会主义道路大步向前，坚持全面深化改革，坚持高质量发展，坚持扩大对外开放，坚持走和平发展道路，推动构建人类命运共同体。

Over the past seven decades, we in China have, based on its realities, constantly explored the way forward through practices, and have succeeded in following the path of socialism with Chinese characteristics. Today, China has reached a new historical starting point. However, we are keenly aware that with all we have achieved, there are still many mountains to scale and many shoals to navigate. We will continue to advance along the path of socialism with Chinese characteristics, deepen sweeping reforms, pursue quality development, and expand opening-up. We remain committed to peaceful development and will endeavor to build a community with a shared future for mankind.

下一步，中国将采取一系列重大改革开放举措，加强制度性、结构性安排，促进更高水平对外开放。

Going forward, China will take a series of major reform and opening-up measures and make stronger institutional and structural moves to boost higher quality opening-up.

第一，更广领域扩大外资市场准入。公平竞争能够提高效率、带来繁荣。中国已实施准入前国民待遇加负面清单管理模式，未来将继续大幅缩减负面清单，推动现代服务业、制造业、农业全方位对外开放，并在更多领域允许外资控股或独资经营。

First, we will expand market access for foreign investment in more areas. Fair competition boosts business performance and creates prosperity. China has already adopted a management model based on pre-establishment national treatment and negative list, and will continue to significantly shorten the negative list. We will work for the all-round opening-up of modern services, manufacturing and agriculture, and will allow the operation of foreign-controlled or wholly foreign-owned businesses in more sectors.

我们将新布局一批自由贸易试验区，加快探索建设自由贸易港。我们将加快制定配套法规，确保严格实施《外商投资法》。我们将以公平竞争、开放合作推动国内供给侧结构性改革，有效淘汰落后和过剩产能，提高供给体系质量和效率。

We will plan new pilot free trade zones and explore at a faster pace the opening of a free trade port. We will accelerate the adoption of supporting regulations to ensure full implementation of the "Foreign Investment Law". We will promote supply-side structural reform through fair competition and open cooperation, and will phase out backward and excessive production capacity in an effective way to improve the quality and efficiency of supply.

第二，更大力度加强知识产权保护国际合作。没有创新就没有进步。加强知识产权保护，不仅是维护内外资企业合法权益的需要，更是推进创新型国家建设、推动高质量发展的内在要求。

Second, we will intensify efforts to enhance international cooperation in intellectual property protection. Without innovation, there will be no progress. Full intellectual property protection will not only ensure the lawful rights and interests of Chinese and foreign companies; it is also crucial to promoting China's innovation-driven and quality development.

中国将着力营造尊重知识价值的营商环境，全面完善知识产权保护法律体系，大力强化执法，加强对外国知识产权人合法权益的保护，杜绝强制技术转让，完善商业秘密保护，依法严厉打击知识产权侵权行为。中国愿同世界各国加强知识产权保护合作，创造良好创新生态环境，推动同各国在市场化法治化原则基础上开展技术交流合作。

China will spare no effort to foster a business environment that respects the value of knowledge. We will fully improve the legal framework for protecting intellectual property, step up law enforcement, enhance protection of the lawful rights and interests of foreign intellectual property owners, stop forced technology transfer, improve protection of trade secrets, and crack down hard on violations of intellectual property in accordance with law. China will strengthen cooperation with other countries in intellectual property protection, create an enabling environment for innovation and promote technological exchanges and cooperation with other countries on the basis of market principles and the rule of law.

第三，更大规模增加商品和服务进口。中国既是"世界工厂"，也是"世界市场"。中国有世界上规模最大、成长最快的中等收入群体，消费增长潜力巨大。为满足人民日益增长的物质文化生活需要，增加消费者选择和福利，我们将进一步降低关税水平，消除各种非关税壁垒，不断开大中国市场大门，欢迎来自世界各国的高质量产品。我们不刻意追求贸易顺差，愿意进口更多国外有竞争力的优质农产品、制成品和服务，促进贸易平衡发展。

Third, we will increase the import of goods and services on an even larger scale. China is both a global factory and a global market. With the world's largest and fastest growing middle-income population, China has a vast potential for increasing consumption. To meet our people's ever-growing material and cultural needs and give our consumers more choices and benefits, we will further lower tariffs and remove various non-tariff barriers. We will steadily open China's market wider to quality products from across the world. China does not seek trade surplus; we want to import more competitive quality agricultural products, manufactured goods and services to promote balanced trade.

第四，更加有效实施国际宏观经济政策协调。全球化的经济需要全球化的治理。中国将加强同世界各主要经济体的宏观政策协调，努力创造正面外溢效应，共同促进世界经济强劲、可持续、平衡、包容增长。中国不搞以邻为壑的汇率贬值，将不断完善人民币汇率形成机制，使市场在资源配置中起决定性作用，保持人民币汇率在合理均衡水平上的基本稳定，促进世界经济稳定。规则和信用是国际治理体系有效运转的基石，也是国际经贸关系发展的前提。中国积极支持和参与世贸组织改革，共同构建更高水平的国际经贸规则。

Fourth, we will more effectively engage in international macro-economic policy coordination. A globalized economy calls for global governance. China will strengthen macro policy coordination with other major economies to generate a positive spillover and jointly contribute to robust, sustainable, balanced and inclusive global growth. China will not resort to the beggar-thy-neighbor practice of RMB devaluation. On the contrary, we will continue to improve the exchange rate regime, see that the market plays a decisive role in resource allocation and keep the RMB exchange rate basically stable at an adaptive and equilibrium level. These steps will help ensure the steady growth of the global economy. Rules and credibility underpin the effective functioning of the international governance system; they are the prerequisite for growing international economic and trade relations. China is an active supporter and participant of WTO reform and will work with others to develop international economic and trade rules of higher standards.

第五，更加重视对外开放政策贯彻落实。中国人历来讲求"一诺千金"。我们高度重视履行同各国达成的多边和双边经贸协议，加强法治政府、诚信政府建设，建立有约束的国际协议履约执行机制，按照扩大开放的需要修改完善法律法规，在行政许可、市场监管等方面规范各级政府行为，清理废除妨碍公平竞争、扭曲市场的不合理规定、补贴和做法，公平对待所有企业和经营者，完善市场化、法治化、便利化的营商环境。

Fifth, we will work harder to ensure the implementation of opening-up related policies. We Chinese have a saying that honoring a promise carries the weight of gold. We are committed to implementing multilateral and bilateral economic and trade agreements reached with other countries. We will strengthen the building of a government based on the rule of law and good faith. A binding mechanism for honoring international agreements will be put in place. Laws and regulations will be revised and improved in keeping with the need to expand opening-up. We will see that governments at all levels operate in a well-regulated way when it comes to issuing administrative licenses and conducting market oversight. We will overhaul and abolish unjustified regulations, subsidies and practices that impede fair competition and distort the market. We will treat all enterprises and business entities equally, and foster an enabling business environment based on market operation and governed by law.

中国扩大开放的举措,是根据中国改革发展客观需要作出的自主选择,这有利于推动经济高质量发展,有利于满足人民对美好生活的向往,有利于世界和平、稳定、发展。我们也希望世界各国创造良好投资环境,平等对待中国企业、留学生和学者,为他们正常开展国际交流合作活动提供公平友善的环境。我们坚信,一个更加开放的中国,将同世界形成更加良性的互动,带来更加进步和繁荣的中国和世界。

These measures to expand opening-up are a choice China has made by itself to advance its reform and development. It will promote high-quality economic development, meet the people's desire for a better life, and contribute to world peace, stability and development. We hope that other countries will also create an enabling environment of investment, treat Chinese enterprises, students and scholars as equals, and provide a fair and friendly environment for them to engage in international exchanges and cooperation. We are convinced that a more open China will further integrate itself into the world and deliver greater progress and prosperity for both China and the world at large.

同事们、朋友们!
Dear Colleagues and Friends,

让我们携起手来,一起播撒合作的种子,共同收获发展的果实,让各国人民更加幸福,让世界更加美好!

Let us join hands to sow the seeds of cooperation, harvest the fruits of development, bring greater happiness to our people and make our world a better place for all!

祝本次高峰论坛圆满成功！

In conclusion, I wish the Second Belt and Road Forum for International Cooperation a full success!

谢谢大家。

Thank you!

阅读拓展
More Reading for Practice

携手推进"一带一路"建设
Work Together
to Build the Silk Road Economic Belt
and The 21st Century Maritime Silk Road

内容参见二维码

练习
Exercises

练习一　翻译课文内容

练习二　阅读理解

While still in its early stages, welfare reform has already been judged a great success in many states—at least in getting people off welfare. It's estimated that more than 2 million people have left the rolls since 1994.

In the past four years, welfare rolls in Athens County have been cut in half. But 70 percent of the people who left in the past two years took jobs that paid less than $6 an hour. The result: The Athens County poverty rate still remains at more than 30 percent—twice the national average.

For advocates for the poor, that's an indication much more needs to be done.

"More people are getting jobs, but it's not making their lives any better," says Kathy Lairn, a policy analyst at the Center on Budget and Policy Priorities in Washington.

A center analysis of US Census data nationwide found that between 1995 and 1996, a greater percentage of single, female-headed households were earning money on their own, but that average income for these households actually went down.

But for many, the fact that poor people are able to support themselves almost as well without government aid as they did with it is in itself a huge victory.

"Welfare was a poison. It was a toxin that was poisoning the family," says Robert Rector, a welfare-reform policy analyst. "The reform is chafing the moral climate in low-income communities. It's beginning to rebuild the work ethnic, which is much more important."

Mr. Rector and others argued that once "the habit of dependency is cracked," then the country can make other policy changes aimed at improving living standards.

1. From the passage, it can be seen that the author _____.
A. believes the reform has reduced the government's burden
B. insists that welfare reform is doing little good for the poor
C. is overenthusiastic about the success of welfare reform
D. considers welfare reform to be fundamentally successful

2. Why aren't people enjoying better lives when they have jobs?
A. Because many families are divorced.
B. Because government aid is now rare.
C. Because their wages are low.

D. Because tile cost of living is rising.

3. What is worth noting from the example of Athens County is that _____.

A. greater efforts should be made to improve people's living standards

B. 70 percent of the people there have been employed for two years

C. 50 percent of the population no longer relies on welfare

D. the living standards of most people are going down

4. From the passage we know that welfare reform aims at _____.

A. saving welfare funds   B. rebuilding the work ethic

C. providing more jobs   D. cutting government expenses

5. According to the passage, before the welfare reform was carded out, _____.

A. the poverty rate was lower

B. average living standards were higher

C. the average worker was paid higher wages

D. the poor used to rely on government aid

第四部分 文化经典

PART FOUR CULTURAL CLASSICS

第一单元 农舍（节选）

UNIT ONE
ON COTTAGES IN GENERAL (EXCERPTED)

For it is not the large houses that live in the memory of the visitor. He goes through them as a matter of duty, and forgets about them as a matter of course. The pictures that linger in his mind, called up in a moment by such sensations as the smell of roses or of new-mown hay, are of a simpler nature. A little cottage nestling amidst the wayside trees, the blue smoke curling up against the green, and a bower of roses round the door; or perhaps a village street of which the name has been long forgotten, with its rambling old inn, and, a little distance away, the hoary, grey church-tower in its township of tombstones—these are the pictures of old England that are carried away to other climes. And it is the cottage, more homely than the inn, more sacred than the church, that we remember best.

Such places have no history at all, their life has not been set in the public eye, and they have always been so wrapt up in their own affairs, that they have never noticed how time is passing, and so they have brought down into the life of today the traditions of two or three hundred years ago.

But though they do not pose, those quiet places, yet it is through them that the deep, main current of English life has flowed. For it is a shallow theory that views history as the annals of a court, or the record of the lives of a few famous men. Doubtless such have their significance, but it is easy to overrate their importance, and they afford but little clue to the life of the people, which is the real history of the country. And until recent days it was not through the cities that this main stream flowed, but through innumerable little country towns and villages.

Washington Irving grasped this fact nearly a hundred years ago when he wrote: "The who stranger would form a correct opinion of English character must go forth into the country. He must sojourn in villages and hamlets; he must visit castles, villas, farmhouses, cottages; he must wander through parks and gardens, along hedges and green lanes; he must loiter about country churches, attend wakes and fairs and other rural festivals, and cope with the people in all their conditions, and all their habits and humors."

And these little villages and hamlets are planted all over England, sometimes close together, sometimes more widely spread, but seldom more than a mile or two apart. Written history may have nothing to say regarding them, but they have helped to make history. They have gathered few legends beyond those which time has written on the walls in weather stains and grey lichen, but the men who were born in those humble cottages have wrought in other lands legends that live today. Their cosy homes were bit newly built when the desperate tides of the civil war surged round them. Half a century later they formed part of the army which "swore terribly in Flanders," and in fifty years more they were laying the foundations of our great Indian empire. Then the arid fields of Spain saw them as they followed the Iron Due through the dogged ears of the Peninsular war, and they took part in his crowning triumph at Waterloo. Later still, India knew them once more, and the snowy trenches of the Crimea, and but yesterday Afghanistan, Egypt, and South Africa called them forth again.

And all the while that those truant birds upheld the name of England abroad, leaving their bones in many lands, their brothers and sisters carried forward the old traditions at home, living their busy, unobtrusive, useful lives, and lying down to rest at last in the old familiar churchyard. And after all, this last is the real life of England, for the sake of which those wars were waged and bloody battles fought. It is the productive life which brings wealth and prosperity and happiness to a nation, and lays the foundation of all that is its honor and its pride.

There is nothing obtrusive about the old cottages. They do not dominate the landscape, but are content to be part of it, and to pass unnoticed unless one looks specially for their homely beauties. The modern house, on the other hand, makes a bid for your notice. It is built on high ground, commands a wide range country, and is seen from far and wide. But the old cottage prefers to nestle snugly in shady valleys. The trees grow closely about it in an intimate, familiar way, and at a little distance only the wreath of curling smoke tells of its presence.

Indeed the old cottage has always been something so very close and so familiar to us, that its charms have been almost entirely overlooked, and it is only of recent years, when fast falling into decay, that it has formed a theme for pen and pencil. Truth to tell, of late years a change has come over England. The life that the old cottage typifies is now a thing of the past, and is daily fading more and more into the distance. Twentieth-century England, the England of the railway, the telegraph, and the motorcar, is not the England of these old cottages. Our point of view has changed. We no longer see the old homely life from within, but from the outside. But the commonplace of yesterday becomes the poetry of today, such a glamour does the magician Time cast over things, and the old life becomes ever more and more attractive as it slips away from us, and we watch it disappear with regretful and kindly eyes.

注解
Notes

本文摘自《英国的农舍》一书。该书由海伦·阿林厄姆及斯图尔特·迪克合著。海伦·阿林厄姆是位出色的画家。她酷爱英国乡村中小小的农舍，善于将恬静朴素的乡村生活情调溶于水彩画中。对她而言，农舍不是没有生命的物体，而是历史的见证人，其外貌与结构体现了不同的历史风貌。海伦·阿林厄姆的农舍画为她赢得了不朽的名声。斯图尔特·迪克是位优秀的画家兼作家。他有丰富的历史知识及深邃的洞察力，对海伦的农舍画理解极为深刻。他以朴素优美的语言描述着幢幢农舍，带领读者去倾听历史的脚步声，去回味已逝去的文明。这部别具一格的著作画面优美动人，文字简洁清新，那含蓄的魅力使其出版后便成为深受读者喜爱的畅销书。

译文参考
Translation for Reference

农舍

（节选）

海伦·阿林厄姆　斯图尔特·迪克

宽大的住宅并不能长久地留在游人的记忆中。他像履行义务似的观赏这些住宅，又极其自然地将它们忘得一干二净。萦绕在他心头的画面，一闻到蔷薇花香或新鲜干草的气息便会立刻在记忆中复苏的情景，其实十分平凡、自然。小小的农舍偎依在路旁的树丛中，绿色的屏障上，青烟袅袅，沿门攀缘的蔷薇，投下一片阴凉；或许，一条不知名的小街上，有座设计不太规范的客店，不远处，灰色教堂的尖顶耸立在乡村墓地中——这才是古老英国的真实写照，这才是英国留给游客的印象。农舍比客店更亲切，比教堂更神圣，它才是深深铭刻在人们心中的英国标记。

农舍这样的地方没有多少历史可言，农舍的生活也鲜为世人所知，它们只埋头于自己的事务，全然不注意时间在如何流逝。就这样，它们把二三百年前的传统保留了下来。

尽管它们无声无息，毫不装模作样，深沉的、英国生活的主流却在那儿流淌。认为历史只是宫廷生活的记录或是寥若晨星的名人传记，当属浅薄之见。毫无疑问，这些记录或传记的确重要，但人们易于过分强调它们的重要性，况且，它们几乎极少提供了解人民生活的线索，殊不知人民的生活才是英国真正的历史，直到不久以前，英国生活的主流还不在城市，而是在乡村，在无计其数的村镇中。

华盛顿·欧文在近百年前就把握住了这一事实。他写道："外国人若要对英国人的性格有恰如其分的了解，就必须到乡村去。他应该在乡村小住数日，参观城堡、别墅、农场房屋及村民居住的农舍，逛逛公园、花园，沿着矮树丛及林荫小路蹓跶；他一定要在教堂里消磨一下时光，参加纪念守护神节日的活动，看看集市，与村民同庆他们的节日，与不同场合的人打交道，了解他们的习惯并欣赏他们的幽默。"

大大小小的村庄，遍布英国各地。有时，几个村庄连成一片，有时又分散在田野间，但彼此间隔极少超过一二英里。有文字记载的历史可能从未提及它们，它们却为创造历史贡献了力量。虽然农舍的墙上只有风雨剥蚀留下的印记及灰色的地衣，找不出动听的故事，诞生在这些寒舍中的人却在异国他乡出了名，其业绩传颂至今。他们刚建好小巧舒适的新家，便卷入了疯狂的内战漩涡。半个世纪以后，他们又加入部队，在"佛兰德信誓旦旦"。五十多年时间内，他们为大印度帝国奠定了基石。接着，西班牙干枯的田野目睹他们跟着"铁公爵"，征战伊比利亚半岛，历尽艰辛；还看到他们分享滑铁卢辉煌胜利的喜悦。后来，他们再次光临印度，克里米亚雪地的战壕也一睹他们的雄风；昨天，仅仅在昨天，阿富汗、埃及与南非便又在召唤他们了。

当这些无暇顾及家室的人在国外为英国扬名，将尸骨抛在异乡时，他们的兄弟姐妹却一直在国内保持着古老的传统。他们忙忙碌碌，不引人注目，其劳动却有益于英国的发展；最后，他们长眠在古老熟悉的教堂墓地中。归根结底，这才是真正的英国生活，为了它，人们才去发动战争，才会血染疆场。正是这种富有创造性的生活，才给英国民族带来了财富、繁荣和幸福，形成了英国的荣耀和骄傲所依附的基础。

古老的农舍丝毫不引人注目，它们并不独占风光，能点缀周围的风景也十分满足；人们只是为欣赏质朴无华的美时才会注意到它们，对此，它们也无怨言。而现代房屋则是千方百计想赢得人们的青睐；它建在高处，可将广阔的乡野尽收眼底，远近也只数它最瞩目。但古老的农舍宁肯掩映在绿莹莹的山谷中，被树木亲密地环抱着，走近了才能看见，否则，便只有从缕缕烟圈才能判断它的所在了。

古老的农舍一直使我们感到太亲近、太熟悉以至对它的魅力熟视无睹。只是到最近它们迅速地崩塌之时，才成为作家和画家作品的主题。的确，英国在变。古老农舍代表的生活已成为历史，离我们一天比一天远。20世纪的英国，火车、电报及汽车的英国已不再是古老农舍的英国了。我们的观点也在变，我们不再从内部而是从外部来看旧时的家庭生活。昔日平凡的事物成为今日诗歌的主题。时间这位魔术师给历史赋予这样的魅力：在它悄悄逝去时，反倒更加迷人。我们只能不无遗憾地、友好地目送着它从视野中消逝。

第四部分 文化经典
PART FOUR CULTURAL CLASSICS

阅读拓展
More Reading for Practice

The Song of the River
W. S. Maugham

You hear it along the river. You hear it, loud and strong, from the rowers as they urge the junk with its high stern, the mast lashed alongside, down the swift running stream. You hear it from the trackers, a more breathless chant, as they pull desperately against the current, half a dozen of them perhaps if they are taking up a wupan, a couple of hundred if they are hauling a splendid junk, its square sails set, over a rapid. On the junk a man stands amid ships, beating a drum incessantly to guide their efforts, and they pull with all their strength, like men possessed, bent double; and sometimes in the extremity of their travail they crawl on the ground, on all fours, like the beasts of the field. They strain, strain fiercely, against the pitiless might of the stream. The leader goes up and down the line and when he sees one who is not putting all his will into the task he brings down his split bamboo on the naked back. Each one must do his utmost or the labor of all is vain. And still they sing a vehement. eager chant, the chant of the turbulent waters. I do not know how words can describe what there is in it of effort. It serves to express the straining heart, the breaking muscles, and at the same time the indomitable spirit of man which overcomes the pitiless force of nature. Though the rope may part and the great junk swing back, in the end the rapid will be passed; and at the close of the weary day there is the hearty meal…

But the most agonizing song is the song of the coolies who bring the great bales from the junk up the steep steps to the town wall. Up and down they go, endlessly, and endless as their toil rises their rhythmic cry. He, aw-ah, oh. They are barefoot and naked to the waist. The sweat pours down their faces and their song is a groan of pain. It is a sigh of despair. It is heart-rending. It is hardly human. It is the cry of souls in infinite distress, only just musical, and that last note is the ultimate sob of humanity. Life is too hard, too cruel, and this is the final despairing protest. That is the song of the river.

注解
Notes

本文是一篇写景抒情的散文。作者毛姆是当代英国著名小说家。他于 20 世纪 20 年代末曾到中国，并有机会沿长江溯流而上进入四川。本文后半部所写可能就是山城重庆的景象。毛姆把他在中国旅行所见写成一本散文集，名为《在中国屏幕上》（On a Chinese screen），本文即选自该书。文章情景交融，读来亲切感人。翻译这样的文章并不像人们初看的那样容易，关键是汉语对应词的选择一定要切合具体情况。

1. wupan 是按汉语"五板"和英语 sampan（舢板）模式造出来的词，指中国的一种小木船。
2. square sail（航海词汇）横帆。
3. beasts 不可译"野兽"，这里指"牲口"。
4. hearty 与 meal 连用时意为 large，即只指"量"而不指"质"。
5. "endless as their toil rises their rhythmic cry" 是个倒装句，主语是 cry。动词 rises 在这样的上下文里不是指"升起"或"提高"而是"发出"。musical 不可译成"音乐的"，这里主要指其节奏。

译文参考
Translation for Reference

河之歌
W. S. 毛姆

沿河上下都可以听见歌声。它响亮而有力，那是船夫，他们划着木船顺流而下，船尾翘得很高，桅杆系在船边。它也可能是比较急促的号子，那是纤夫，他们拉纤逆流而上，如果拉的是小木船，也许就只五六个人，如果拉的是扬着横帆的大船过急滩，那就要二百来人。船中央站着一个汉子不停地击鼓助威，引导他们加劲。于是他们使出全部力量，像着了魔似的，腰弯成两折，有时力量用到极限就全身趴在地上匍匐前进，像田里的牲口。他们使劲，拼命使劲，对抗着水流无情的威力。领头的在纤绳前后跑来跑去，见到有人没有全力以赴，竹板就打在他光着的背上。每个人都必须竭尽全力，否则就会前功尽弃。就这样他们还是唱着激昂而热切的号子，那汹涌澎湃的河水号子。我不知道词语怎样能描写出其中所包含的拼搏，它表现的是绷紧的心弦，几乎要断裂的筋肉，同时也表现了人类克服无情的自然力的顽强精神。虽然绳子可能扯断，大船可能倒退，但最终险滩必将通过，在筋疲力尽的一天结束时可以痛快地吃上一顿饱饭。

然而最令人难受的却是苦力的歌，他们背负着船上卸下的大包，沿着陡坡爬上城墙。他们不停地上上下下，随着无尽的劳动响起有节奏的喊声：嗨，哟——嗬，嗨。他们赤着脚，光着背，汗水不断地从脸上流下。他们的歌是痛苦的呻吟，失望的叹息，听来令人心碎，简直不像是人的声音。它是灵魂在无尽悲戚中的呼喊，只不过有着音乐的节奏而已。那终了的一声简直就是人性泯灭的低泣。生活太艰难，太残酷，这喊声正是最后绝望的抗议。这就是河之歌。

练习
Exercises

练习一　翻译原课文

练习二　阅读理解

Passage A

Scientists have established that influenza viruses taken from man can cause the disease in animals. In addition, man can catch the disease from animals. In fact, a great number of wild birds seem to carry the virus without showing any evidence of illness. Some scientists conclude that a large family of influenza viruses may have evolved in the bird kingdom, a group that has been on the earth 100 million years and is able to carry the virus without contracting the disease. There is even convincing evidence to show that virus strains are transmitted from place to place and from continent to continent by migrating birds. It is known that two influenza viruses can recombine when both are present in an animal at the same time. The result of such recombinations is a great variety of strains containing different H and N spikes. This raises the possibility that a human influenza virus can recombine with an influenza virus from a lower animal to produce an entirely new spike. Research is underway to determine if that is the way that major new strains come into being. Another possibility is that two animal influenza strains may recombine in a pig, for example, to produce a new strain which is transmitted to man.

1. According to the passage, scientists have discovered that influenza viruses _____.

   A. cause ill health in wild birds

   B. do not always cause symptoms in birds

   C. are rarely present in wild birds

   D. change when transferred from animals to man

2. What is known about the influenza virus?

   A. It was first found in a group of very old birds.

   B. All the different strains can be found in wild birds.

   C. It existed over 100 million years ago.

   D. It can survive in many different places.

3. According to the passage, a great variety of influenza strains can appear when _____.

   A. H and N spikes are produced

   B. animal and bird viruses are combined

   C. dissimilar types virus recombine

   D. two viruses of the same type are contracted

4. New strains of viruses are transmitted to man by _____.

   A. a type of pig   B. diseased lower animals

   C. a group of migrating birds   D. a variety of means

5. It can be inferred from the passage that all of the following are ways of producing new strains of influenza viruses EXCEPT _____.

   A. two influenza viruses in the same animal recombining

B. animal viruses recombining with human viruses
C. two animal viruses recombining in one animal
D. two animal viruses recombining in a human

Passage B

Water on the earth is being recycled continuously in a process known as the hydrologic cycle. The first step of the cycle is the evaporation of water in the oceans. Evaporation is the process of water turning into vapor, which then forms clouds in the sky. The second step is the water returning to the earth in the form of precipitation: either rain, snow, or ice. When the water reaches the earth's surface, it run off into the rivers, lakes, and the ocean, where the cycle begins again. Not all water, however, stays on the surface of the earth in the hydrologic cycle. Some of it seeps into the ground through infiltration and collects under the earth's surface as groundwater. This groundwater is extremely important to life on earth, since 95 percent of the earth's water is in the oceans and is too salty for human beings or plants. Of the 5 percent on land, only 0.5 percent is above ground in rivers or lakes. The rest is underground water. This groundwater is plentiful and dependable, because it doesn't depend on seasonal rain or snow. It is the major source of water for many cities. But as the population increases and the need for water also increases, the groundwater in some areas is getting dangerously low. Added to this problem is an increasing amount of pollution that seeps into the groundwater. In the future, with a growing population and more toxic water, the hydrologic cycle we depend on could become dangerously imbalanced.

1. Clouds are formed from _____.
   A. water vapor   B. evaporation
   C. the hydrologic cycle   D. groundwater

2. Water returns to the earth by _____.
A. infiltration   B. pollution
C. precipitation   D. evaporation
3. Groundwater _____.
A. depends on seasonal rain   B. comes from toxic waste
C. is 0.5 percent of all water   D. collects under the earth
4. The amount of groundwater is _____.
A. about 95 percent of all water
B. less than 5 percent of all water
C. 0.5 percent of above-ground water
D. 95 percent of above-ground water
5. The supply of groundwater is going low because of _____.
A. conservation   B. toxic waste
C. pollution   D. population increase

第二单元 荷塘月色
UNIT TWO
MOONLIGHT OVER THE LOTUS POND

这几天心里颇不宁静。今晚在院子里坐着乘凉，忽然想起日日走过的荷塘，在这满月的光里，总该另有一番样子吧。月亮渐渐地升高了，墙外马路上孩子们的欢笑，已经听不见了；妻在屋里拍着闰儿，迷迷糊糊地哼着眠歌。我悄悄地披了大衫，带上门出去。

I have felt quite upset recently. (It has been rather disquieting these days.) Tonight, when I was sitting in the yard enjoying the cool, it occurred to me that the Lotus Pond, which I pass by every day, must assume quite a different look in such moonlit night. A full moon was rising high in the sky; the laughter of children playing outside had died away; in the room, my wife was patting the son, Run-er, sleepily humming a cradle song. Shrugging on an overcoat, quietly, I made my way out, closing the door behind me.

沿着荷塘，是一条曲折的小煤屑路。这是一条幽僻的路；白天也少人走，夜晚更加寂寞。荷塘四面，长着许多树，蓊蓊郁郁的。路的一旁，是些杨柳，和一些不知道名字的树。没有月光的晚上，这路上阴森森的，有些怕人。今晚却很好，虽然月光也还是淡淡的。

Alongside the Lotus Pond runs a small cinder footpath. It is peaceful and secluded here, a place not frequented by pedestrians even in the daytime; now at night, it looks more solitary, in a lush, shady ambience of trees all around the pond. On the side where the path is, there are willows, interlaced with some others whose names I do not know. The foliage, which, in a moonless night, would loom somewhat frighteningly dark, looks very nice tonight, although the moonlight is not more than a thin, grayish veil.

路上只我一个人，背着手踱着。这一片天地好像是我的；我也像超出了平常的自己，到了另一世界里。我爱热闹，也爱冷静；爱群居，也爱独处。像今晚上，一个人在这苍茫的月下，什么都可以想，什么都可以不想，便觉是个自由的人。白天里一定要做的事，一定要说的话，现在都可不理，这是独处的妙处。我且受用这无边的荷香月色好了。

I am on my own, strolling, hands behind my back. This bit of the universe seems in my possession now; and I myself seem to have been uplifted from my ordinary self into another world. I like a serene and peaceful life, as much as a busy and active one; I like being in solitude, as much as in company. As it is tonight, basking in a misty moonshine all by myself, I feel I am a free man, free to think of anything, or of nothing. All that one is obliged to do, or to say, in the daytime, can be very well cast aside now. That is the beauty of being alone. For the moment, just let me indulge in this profusion of moonlight and lotus fragrance.

曲曲折折的荷塘上面，弥望的是田田的叶子。叶子出水很高，像亭亭的舞女的裙。层层的叶子中间，零星地点缀着些白花，有袅娜地开着的，有羞涩地打着朵儿的；正如一粒粒的明珠，又如碧天里的星星，又如刚出浴的美人。微风过处，送来缕缕清香，仿佛远处高楼上渺茫的歌声似的。这时候叶子与花也有一丝的颤动，像闪电般，霎时传过荷塘的那边去了。叶子本是肩并肩密密地挨着，这便宛然有了一道凝碧的波痕。叶子底下是脉脉的流水，遮住了，不能见一些颜色；而叶子却更见风致了。

All over this winding stretch of water, what meets the eye is a silken field of leaves, reaching rather high above the surface, like the skirts of dancing girls in all their grace. Here and there, layers of leaves are dotted with white lotus blossoms, some in demure bloom, others in shy bud, like scattering pearls, or twinkling stars, or beauties just out of the bath. A breeze stirs, sending over breaths of fragrance, like faint singing drifting from a distant building. At this moment, a tiny thrill shoots through the leaves and lilies, like, a streak of lightning, straight across the forest of lotuses. The leaves, which have been standing shoulder to shoulder, are caught shimmering in an emerald heave of the pond. Underneath, the exquisite water is covered from view, and none can tell its colour; yet the leaves on top project themselves all the more attractively.

月光如流水一般,静静地泻在这一片叶子和花上。薄薄的青雾浮起在荷塘里。叶子和花仿佛在牛乳中洗过一样,又像笼着轻纱的梦。虽然是满月,天上却有一层淡淡的云,所以不能朗照;但我以为这恰是到了好处——酣眠固不可少,小睡也别有风味的。月光是隔了树照过来的,高处丛生的灌木,落下参差的斑驳的黑影,峭楞楞如鬼一般;弯弯的杨柳的稀疏的倩影,却又像是画在荷叶上。塘中的月色并不均匀,但光与影有着和谐的旋律,如梵婀铃上奏着的名曲。

The moon sheds her liquid light silently over the leaves and flowers, which, in the floating transparency of a bluish haze from the pond, look as if they had just been bathed in milk, or like a dream wrapped in a gauzy hood. Although it is a full moon, shining through a film of clouds, the light is not at its brightest; it is, however, just right for me-a profound sleep is indispensable, yet a snatched doze also has a savour of its own. The moonlight is streaming down through the foliage, casting bushy shadows on the ground from high above, jagged and checkered, as grotesque as a party of spectres; whereas the benign figures of the drooping willows, here and there, look like paintings on the lotus leaves. The moonlight is not spread evenly over the pond, but rather in a harmonious rhythm of light and shade, like a famous melody played on a violin.

荷塘的四面,远远近近,高高低低都是树,而杨柳最多。这些树将一片荷塘重重围住;只在小路一旁,漏着几段空隙,像是特为月光留下的。树色一例是阴阴的,乍看像一团烟雾;但杨柳的丰姿,便在烟雾里也辨得出。树梢上隐隐约约的是一带远山,只有些大意罢了。树缝里也漏着一两点路灯光,没精打采的,是瞌睡人的眼。这时候最热闹的,要数树上的蝉声与水里的蛙声;但热闹是它们的,我什么也没有。

Around the pond, far and near, high and low, are trees. Most of them are willows. Only on the path side, can two or three gaps be seen through the heavy fringe, as if specially reserved for the moon. The shadowy shapes of the leafage at first sight seem diffused into a mass of mist, against which, however, the charm of those willow trees is still discernible. Over the trees appear some distant mountains, but merely in sketchy silhouette. Through the branches are also a couple of lamps, as listless as sleepy eyes. The most lively creatures here, for the moment, must he the cicadas in the trees and the frogs in the pond. But the liveliness is theirs, I have nothing.

忽然想起采莲的事情来了。采莲是江南的旧俗，似乎很早就有，而六朝时为盛，从诗歌里可以约略知道。采莲的是少年的女子，她们是荡着小船，唱着艳歌去的。采莲人不用说很多，还有看采莲的人。那是一个热闹的季节，也是一个风流的季节。梁元帝《采莲赋》里说得好：

Suddenly, something like lotus-gathering crosses my mind. It used to he celebrated as a folk festival in the South, probably dating very far hack in history, mast popular in the period of Six Dynasties. We can pick up some outlines of this activity in the poetry. It was young girls who went gathering lotuses, in sampans and singing love songs. Needless to say, there were a great number of them doing the gathering, apart from those who were watching. It was a lively season, brimming with vitality, and romance. A brilliant description can be found in "Lotus Gathering" written by the Yuan Emperor of the Liang Dynasty：

于是妖童媛女，荡舟心许：鷁首徐回，兼传羽杯；棹将移而藻挂，船欲动而萍开。尔其纤腰束素，迁延顾步；夏始春余，叶嫩花初，恐沾裳而浅笑，畏倾船而敛裾。

So those charming youngsters row their sampans, heart buoyant with tacit love, pass on to each other cups of wine while their bird-shaped prows drift around. From time to time their oars are caught in dangling algae, and duckweed flow apart the moment their boats are about to move on. Their slender figures, girdled with plain silk, tread watchfully on board. This is the time when spring is growing into summer, the leaves a tender green and the flowers blooming-among which the girls are giggling when evading an out-reaching stem, their skirts tucked in for fear that the sampan might tilt.

可见当时嬉游的光景了。这真是有趣的事，可惜我们现在早已无福消受了。于是又记起《西洲曲》里的句子：

That is a glimpse of those merrymaking scenes. It must have been fascinating: but unfortunately we have long been denied such a delight. Then I recall those lines in "Ballad of Xizhou Island":

采莲南塘秋，莲花过人头；低头弄莲子，莲子清如水。

Gathering the lotus, I am in the South Pond, / The lilies in autumn reach over my head; / Lowering my head I toy with the lotus seeds. / Look, they are as fresh as the water underneath.

今晚若有采莲人，这儿的莲花也算得"过人头"了；只不见一些流水的影子，是不行的。这令我到底惦着江南了。——这样想着，猛一抬头，不觉已是自己的门前；轻轻地推门进去，什么声息也没有，妻已睡熟好久了。

If there were somebody gathering lotuses tonight, she could tell that the lilies here are high enough to "reach over her head"; but, one would certainly miss the sight of the water. So my memories drift back to the South after all. Deep in my thoughts, I looked up, just to find myself at the door of my own house. Gently I pushed the door open and walked in. Not a sound inside, my wife had been fast asleep for quite a while.

1927 年 7 月，北京清华园

Qinghua Campus, Beijing July, 1927

练习
Exercises

练习一　笔译或视译课文

练习二　阅读理解

Passage A

We can see how the product life cycle works by looking at the introduction of instant coffee. When it was introduced, most people did not like it as well as "regular" coffee, and it took several years to gain general acceptance (introduction stage). At one point, though, instant coffee grew rapidly in popularity, and many brands were introduced (stage of rapid growth). After a while, people became attached to one brand and sales leveled off (stage of maturity). Sales went into a slight decline when freeze-dried coffees were introduced (stage of decline).

The importance of the product life cycle to marketers is this: Different stages in the product life cycle call for different strategies. The goal is to extend product life so that sales and profits de not decline. One strategy is called market modification. It means that marketing managers look for new users and market sections. Did you know, for example, that the backpacks that so many students carry today were originally designed for the military?

Market modification also means searching for increased usage among present customers or going for a different market, such as senior citizens. A marketer may re-position the product to appeal to new market sections.

Another product extension strategy is called product modification. It involves changing product quality, features, or style to attract new users or more usage from present users. American auto manufacturers are using quality improvement as one way to recapture world markets. Note, also, how auto manufacturers once changed styles dramatically from year to year to keep demand from falling.

1. According to the passage, when people grow fond of one particular brand of a product, its sales will _____.
   A. decrease gradually   B. remain at the same level
   C. become unstable   D. improve enormously

2. The first paragraph tells us that a new product is _____.

A. not easily accepted by the public

B. often inferior to old ones at first

C. often more expensive than old ones

D. usually introduced to satisfy different tastes

3. Marketers need to know which of the four stages a product is in so as to _____.

A. promote its production   B. work out marketing policies

C. speed up its life cycle   D. increase its popularity

4. The author mentions the example of "backpacks" (Line 5, Para. 2) to show the importance of _____.

A. pleasing the young as well as the old

B. increasing usage among students

C. exploring new market sections

D. serving both military and civil needs

5. In order to recover their share of the world market, U.S. auto makers are _____.

A. improving product quality

B. increasing product features

C. modernizing product style

D. re-positioning their product in the market

Passage B

Foxes and farmers have never got on well. These small dog-like animals have long been accused of killing farm animals. They are officially classified as harmful and farmers try to keep their numbers down by shooting or poisoning them.

Farmers can also call on the services of their local hunt to control the fox population. Hunting consists of pursuing a fox across the countryside, with a group of specially trained dogs, followed by men and women riding horses. When the dogs eventually catch the fox, they kill it or a hunter shoots it.

People who take part in hunting think of it as a sport; they wear a special uniform of red coats and white trousers, and follow strict codes of behavior. But owning a horse and hunting regularly is expensive, so most hunters are wealthy.

It is estimated that up to 100,000 people watch or take part in fox hunting. But over the last couple of decades, the number of people opposed to fox hunting, because they think it is brutal, has risen sharply. Nowadays it is rate for a hunt to pass off without some kind of confrontation between hunters and hunt saboteurs. Sometimes these incidents lead to violence, but mostly saboteurs interfere with the hunt by misleading riders and disturbing the trail of the fox's smell, which the dogs follow.

Noisy confrontations between hunters and saboteurs have become so common that they are almost as much a part of hunting as the pursuit of foxes itself. But this year supporters of fox hunting face a much bigger threat to their sport. A Labor Party Member of the Parliament, Mike Foster, is trying to get Parliament to approve a new law which will make the hunting of wild animals with dogs illegal. If the law is passed, wild animals like foxes will be protected under the ban in Britain.

1. Rich people in Britain have been hunting foxes _____.
A. for recreation
B. in the interests of the farmers
C. to limit the fox population
D. to show off their wealth
2. What is special about fox hunting in Britain?
A. It involves the use of a deadly poison.
B. It is a costly event which rarely occurs.
C. The hunters have set rules to follow.
D. The hunters have to go through strict training.

3. Fox hunting opponents often interfere in the game _____.
A. by resorting to violence
B. by confusing the fox hunters
C. by taking legal action
D. by demonstrating on the scene

4. A new law may be passed by the British Parliament to _____.
A. prohibit farmers from hunting foxes
B. forbid hunting foxes with dogs
C. stop hunting wild animals in the countryside
D. prevent largo-scale fox hunting

5. It can be inferred from the passage that _____.
A. killing foxes with poison is illegal
B. limiting the fox population is unnecessary
C. hunting foxes with dogs is considered cruel and violent

第三单元 清明（诗歌）
UNIT THREE
THE QINGMING FESTIVAL (POEM)

　　清明节又叫踏青节，在仲春与暮春之交，一般是在公历 4 月 5 号前后。清明节是农历节气，是中国传统节日，是祭祖和扫墓的日子，距今已有 2500 多年的历史。清明节与端午节、春节、中秋节并称为中国四大传统节日。2006 年 5 月 20 日，中国文化部申报的清明节经国务院批准列入第一批国家级非物质文化遗产名录。

　　《清明》是著名的唐诗，为唐代文学家杜牧所作。此诗描写了清明的春雨情景。杜牧《清明》英译本有多种，不同的译者根据诗歌的意境翻译并创造出富有自身特色的译文。

清　明
〔唐〕杜牧

清明时节雨纷纷，
路上行人欲断魂。
借问酒家何处有？
牧童遥指杏花村。

首先我们来看一下 28 篇不同的译文。

译文 1　On Pure Bright Day

Pure bright season comes with fine fast drizzle.
and travelers on the road feel their souls sliced off.
Please tell me where I can find a wineshop?
A cowherd boy points to distant Apricot Blooming Village.

译文 2　All Souls' Day

The rain falls thick and fast on All Souls' Day,
The men and women sadly move along the way.
They ask where wineshops can be found or where to rest,

And there the herdboy's fingers Almond-Town suggest.

译文 3 The Mourning Day

A fine rain falls on the tomb-sweeping day,
All mourners are heartbroken on their way.
"A wine shop to rid my sorrow, but where?"
A cowherd points Apricot Cot "O'er there".

译文 4 The Qingming Festival

On the Qingming Day, rain is falling uninterrupted,
And the wayfarer is more than ever broken-hearted;
"Is there any wine shop hereabouts?" A buffalo boy
Points to a distant village where apricots are spotted.

译文 5 Qingming Festival

Qingming rains never seem to end.
The traveler along this road is overcome by dejection.
"Where can I find a tavern?"
"Apricot Village, way down the road,"
A cow boy replies pointing his finger.

译文 6 The Qingming Festival

In these memorial days, annoying drizzle falls;
Mourners en route feel the pain in their hearts.
Is there any wine shop around here, please?
A cowkid points at a faraway Village Apricots.

译文 7 The Day of Clear and Bright

Round clear and bright showers are so frequent;
Wayfarers on the road feel despondent.
"Tell me, Buffalo Boy, is there a tavern somewhere?"
The lad pointed to a hamlet with blossoming apricot trees way down the road.

译文 8　Qingming

It is Qingming, early April, a season of mizzles and gloom,
Away from home, a wayfarer, faring into gloom and doom.
O where can be found a tavern, my goodlad, if I may ask?
There! points the herd-boy to a village where apricots bloom.

译文 9　Qingming Day

As the Qingming Festival comes around,
Thick and fast, the rain is falling down.
On the way to the graveyard to pay homage.
I visit my deceased with a broken heart.
Where may I find a wineshop to kill my sorrow?
A cowboy the way to Apricot Flowers Village shows.

译文 10　Ching Ming Festival

Continuous drizzle on Ching Ming Death Festival days
Pedestrians on the roads seemed to be losing balance
When I try to inquire the whereabout of an alehouse
The cowherd points at distant Apricot Flower Village

译文 11　The Qingming Festival

On the day paying homage at mausoleums and touring,
The successive fine rain makes a passer so heartbreaking.
"Do you know is there a wine shop hereabouts?" he enquires,
A shepherd boy shows the huts amid apricot flowers.

译文 12　The Clear-and-Bright Feast

Upon the Clear-and-Bright Feast of spring, the rain drizzles down in spray.
Pedestrians on countryside ways, in gloom are pinning away.
When asked "Where a tavern fair for rest, is hereabouts to be found",
The shepherd boy the Apricot Bloom Vill, does point to afar and say.

PART FOUR　CULTURAL CLASSICS

译文 13　Pure Brightness (Qing Ming) Festival

It's drizzling thickly and profusely
On the Pure Brightness Day.
Away farer is overwhelmed with sorrows
On his way.
"Excuse me, can you tell me
Where to find a wineshop in the village?"
"Over there," the shepherd boy pointing to
The distant Apricot Blossoms Village.

译文 14　Pure Brightness

The rain patters in the term of Pure Brightness
Passengers along the roads are all of unhappiness
I ask where a wineshop is
The shepherd boy points to Xinghua Village

译文 15　The Tomb-Visiting Day

The ceaseless drizzle drips all the dismal day,
So broken-hearted fares the traveler on the way.
When asked where could be found a tavern bower,
A cowboy points to yonder village of the apricot flower.

译文 16　Qingming

It drizzles and drizzles on this Pure Brightness Day;
I feel heavy at heart, a wayfarer on my journey.
When I ask where a tavern might be found,
The cowherd points yonder to a village with flowering apricot trees.

译文 17　The Pure Brightness Day

It drizzles thick and fast on the Pure Brightness Day,
I travel with my heart lost in dismay.

"Is there a public house somewhere, cowboy?"
He points at Apricot Village faraway.

译文 18 The Mourning Day

It drizzles thick and fast on the Mourning Day,
The mourner's heart is going to break on his way.
When asked for a wineshop to drown his sad hours?
A cowboy points to a hamlet amid apricot flowers.

译文 19 Pure Brightness Festival

Around the Pure Brightness Day it drizzles quite often,
Men on the way to mourning the dead seem heart-broken.
"Where," may I ask, "to find an inn to drown my grievance"?
The herd boy points to Apricot Hamlet in the distance.

译文 20 Tomb Sweeping Day

Tomb Sweeping Day sees drizzles running'n flying,
and hearts lost in gloom, mourners on paths crying.
'Any tavern near and far?' I ask a boy,
who points to Almond Bloom Vill beyond eyeing.

译文 21 The Mourning Day

A drizzling rain falls like tears on the Mourning Day;
The mourner's heart is going to break on his way.
Where can a wineshop be found to drown his sad hours?
A cowherd points to a cot' mid apricot flowers.

译文 22 The Qingming Festival

During the Qing-Ming Festival,
Is daily continual rain.
People, going to the country,
Form an almost unbroken train.

"Is there a wine shop hereabouts?"
We enquire of a peasant lad.
"Apricot Flower Village yonder,
Is a shop where wine can be had."

译文 23　In the Raining Season of Spring

It drizzles endless during the rainy season in spring,
Travelers along the road look gloomy and miserable.
When I ask a shepherd boy where I can find a tavern,
He points at a distant hamlet nestling amidst apricot blossoms.

译文 24　Tomb-Sweeping Day

On Tomb-Sweeping Day it rains endlessly;
On the verge of heart-broken are wayfarers.
Where is the wine shop to drown my sorrow?
The cowboy points afar to Apricot Village.

译文 25　Pure Brightness Day

The rain rustles on at Pure Brightness Day;
The traveler's tired, his soul in dismay.
Where is a tavern? He asks a shepherd,
Who points to Apricot Bloom faraway.

译文 26　Clear-Bright Day

It's raining very hard on Clear-Bright Day;
The traveler's heart is surely wrung on the way.
"Please, where do you know of a tavern's wayside cot?"
The cowboy points to flowers of the apricot.

译文 27　In the Rainy Season of Spring

It drizzles endlessly during the rainy season in spring,
Travellers along the roadlook gloomy and miserable.

When I ask a shepherd boy where I can find a tavern,
He points at a distant hamlet nestling amidst apricot blossoms.

译文 28　Qingming Festival

It drizzles and mizzles in Qingming season
And people on road seem to have lost their souls
Where can I find a tavern, I ask
A shepherd, who points to Apricot Hamlet yonder

注解和分析

阅读、翻译过程也是分析过程。下面我们简单分析以上所有译文。

题目"清明"的翻译有"Qingming","Qingming Day","Qingming Festival","The Qingming Festival","Ching Ming Festival","On Pure Bright Day","All Souls' Day","The Mourning Day","The Day of Clear and Bright","The Clear-and-Bright Feast","Pure Brightness","Pure Brightness Festival","Pure Brightness（Qing Ming）Festival","Pure Brightness Day","The Pure Brightness Day","Clear-Bright Day","The Tomb-Visiting Day","Tomb Sweeping Day","The Mourning Day","In the Raining Season of Spring"。现在,"清明"常指"清明节",一般译为"The Qingming Festival"。

该七言绝句诗多种译文各不相同,各有特点。第一句"清明时节雨纷纷"中"雨纷纷"就用了不同的词语。"fine fast drizzle","The rain falls thick and fast","A fine rain falls","rain is falling uninterrupted","rains never seem to end","annoying drizzle falls","showers are so frequent","It's drizzling all the … long","a season of mizzles","Thick and fast, the rain is falling down","Continuous drizzle","The successive fine rain","nonstop misty rain","the rain drizzleth down in spray","It's drizzling thickly and profusely","The rain patters","The ceaseless drizzle drips","It drizzles and drizzles","It drizzles thick and fast","it drizzles quite often","drizzles running and flying","A drizzling rain falls like tears","daily continual rain","It drizzles endless","it rains endlessly","The rain rustles","It's raining very hard","It drizzles endlessly", "It dizzles and mizzles" ……真是大雨小雨都有，轻重缓急各异。也有从落雨的声音来译写清明时节雨纷纷的情景的。无论怎样译写都是体现出译者对原文情景的理解想象和诗意的还原。

下面我们来看看第二句"路上行人欲断魂"中"断魂"的处理。

"feel their souls sliced off", "sadly (move along the way)", "heartbroken (on their way)", "is more than ever broken-hearted", "is overcome by dejection", "feel the pain in their hearts", "feel despondent", "faring into gloom and doom", "with a broken heart", "seemed to be losing balance", "makes a passer so heartbreaking", "look miserable", "in gloom (are pinning away)", "is overwhelmed with sorrows", "are all of unhappiness", "So broken-hearted", "feel heavy at heart", "with my heart lost in dismay", "(The mourner's) heart is going to break", "seem heart-broken", "hearts lost in gloom, mourners on paths crying", "The mourner's heart is going to break", "look gloomy and miserable", "On the verge of heart-broken", "his soul in dismay", "The traveler's heart is surely wrung", "look gloomy and miserable", "seem to have lost their souls" ……可以看出时态的不同、时空的变化。而动词单复数标记也取决于译者逻辑主语"行人"的单复数选用，无论"行人"用何人称或如何表达。汉语名词没有单数复数，也没有冠词，英语则有，译者必须做出选择。"断魂"一词的处理亦然各有不同，每个译者为了所译整个诗歌的"意美""音美""形美"而作出不同的词语选择、表达选择、翻译选择。

第三句，先看"借问酒家何处有"译写的视角方面。有的译者整首诗词翻译视角纯粹置身画外，多用第三人称或如同原文模糊处理；而有的则将第一人称带入其中，其画面感则是即在画外又在其中。"Please tell me where I can find a wineshop"和"They ask where wineshops can be found or where to rest"视角明显不同。直接引语和间接引语的译写方式也涉及视角的不同。行人是"我"，是"他"还是"他们"？"借问"的主体是"我"、是"他"还是"他们"？各译文不尽相同。而类似"Is there any wine shop around here, please"的翻译则保留了原文视角方面的模糊性。

第四句，"牧童遥指杏花村"。牧童的词语也不止一个，译者的选择也不尽相同。而杏花村的翻译则更是多样化，体现出不同的画面感。"Apricot Blooming Village" "Apricot Flowers Village" "Apricot Blossoms

Village"……有杏有花有村且首字母大写;"Almond-Town"没有特意突出花,"village where apricots are spotted"看见的是花还是果则各是不同的色彩;"Apricot Village, way down the road"沿着道路一直下去的村庄;"a faraway Village Apricots"远方的村庄;"a hamlet with blossoming apricot trees way down the road",沿着开着杏花的树的路下去的村庄;"apricot-flower village" "a village where apricots bloom" "the huts amid apricot flowers"杏花正开着的村庄; "a village with flowering apricot trees"有许多正开着杏花的树的村庄;"Villa Apricot"有异国情调的村庄;"Xinghua Village"仅地名而已,那儿有无花、果、树则需想象填补;"yonder village of the apricot flower" "Apricot Hamlet yonder"有怀旧色彩;"a distant hamlet nestling amidst apricot blossoms"远方掩隐在杏花中的小村庄;"flowers of the apricot"只见花不见村,是酒家就在繁花后面吗? 同一首诗的英文译文不尽相同,不同的视角、不同的焦点、不同的时空、不同的色彩……

最后笔者再次试译:

Qingming

It drizzles and mizzles the whole mourning day,
Travellers one after another are stumbling on their way;
Where is the tavern to warm the sole of dismay?
Points the shepherd boy to the Apricot Village faraway.

(中诗网:www.yzs.com 2018-04-22)

第四单元 秋水（节选）

庄子

UNIT FOUR
THE FLOODS OF AUTUMN (EXCERPTED)

Zhuang Zi

秋水时至，百川灌河；泾流之大，两涘渚崖之间，不辨牛马。于是焉河伯欣然自喜，以天下之美为尽在己。顺流而东行，至于北海，东面而视，不见水端，于是焉河伯始旋其面目，望洋向若而叹，曰："野语有之曰，'闻道百，以为莫己若者。'我之谓也。且夫我尝闻少仲尼之闻而轻伯夷之义者，始吾弗信，今我睹子之难穷也，吾非至于子之门则殆矣，吾长见笑于大方之家。"

北海若曰："井蛙不可以语于海者，拘于虚也；夏虫不可以语于冰者，笃于时也；曲士不可以语于道者，束于教也。今尔出于崖涘，观于大海，乃知尔丑，尔将可与语大理矣。天下之水，莫大于海，万川归之，不知何时止而不盈；尾闾泄之，不知何时已而不虚；春秋不变，水旱不知。此其过江河之流，不可为量数。而吾未尝以此自多者，自以比形于天地而受气于阴阳，吾在于天地之间，犹小石小木之在大山也。方存乎见少，又奚以自多！计四海之在天地之间也，不似礨空之在大泽乎？计中国之在海内，不似稊米之在大仓乎？号物之数谓之万，人处一焉；人卒九州，谷食之所生，舟车之所通，人处一焉。此其比万物也，不似毫末之在于马体乎？五帝之所连，三王之所争，仁人之所忧，任士之所劳，尽此矣。伯夷辞之以为名，仲尼语之以为博，此其自多也；不似尔向之自多于水乎？"

河伯曰："然则吾大天地而小豪末，可乎？"北海若曰："否。夫物，量无穷，时无止，分无常，终始无故。是故大知观于远近，故小而不寡，大而不多，知量无穷，证曏今故，故遥而不闷，掇而不跂，知时无止；察乎盈虚，故得而不喜，失而不忧，知分之无常也；明乎坦涂，故生而不说，死而不祸，知终始之不可故也。计人之所知，不若其所不知；其生之时，不若未生之时。以其至小，求穷其至大之域，是故迷乱而不能自得也。由此观之，又何以知毫末之足以定至细之倪！又何以知天地之足以穷至大之域!"

河伯曰:"世之议者皆曰:'至精无形,至大不可围。'是信情乎?"北海若曰:"夫自细视大者不尽,自大视细者不明。夫精,小之微也;垺,大之殷也;故异便。此势之有也。夫精粗者,期于有形者也;无形者,数之所不能分也;不可围者,数之所不能穷也。可以言论者,物之粗也;可以意致者,物之精也;言之所不能论,意之所不能察致者,不期精粗焉。是故大人之行,不出乎害人,不多仁恩;动不为利,不贱门隶;货财弗争,不多辞让;事焉不借人,不多食乎力,不贱贪污;行殊乎俗,不多辟异;为在从众,不贱佞谄;世之爵禄不足以为劝,戮耻不足以为辱;知是非之不可为分,细大之不可为倪。闻曰:'道人不闻,至德不得,大人无己'。约分之至也。"

现代汉语译文和英语译文参考

秋天的雨水按照时令到来了,成百的河流里的水灌到了黄河里,水流的浩大宽广,从两岸小岛水边高地之间来看,分不清是牛是马。在这时河伯内心喜悦不已,认为天下的美景都在自己这儿了。他顺着水流向东行进,到了北海,脸转向东面看去,看不见水边。这时河伯才改变自得的态度,看着海洋,向着海神若叹息,说:"俗语中有这样的话:'听说的道理很多了,认为没有人比得上自己。'说的就是我啊。并且我曾听说过有认为仲尼的见识少和看轻伯夷大义的人,一开始我不相信,现在我看到了您的无穷无尽,我要是没有来到您的门前就危险了,我将长久地被内行笑话了。"

The time of the autumnal floods was come, and the hundred streams were all discharging themselves into the He. Its current was greatly swollen, so that across its channel from bank to bank one could not distinguish an ox from a horse. On this the (spirit-) earl of the He laughed with delight, thinking that all the beauty of the world was to be found in his charge. Along the course of the river he walked east till he came to the North Sea, over which he looked with his face to the east, without being able to see where its waters began. Then he began to turn his face round looked across the expanse, (as if he were) confronting Ruo, and said with a sigh, "What the vulgar saying expresses about him who has learned a hundred points (of the Dao), and thinks that there is no one equal to himself, was surely spoken of me. And moreover, I have heard parties making little of the knowledge of Zhongni and the righteousness of Bo-yi, and at first I did not believe them. Now I behold the all-but-boundless extent (of your realms). If I had not come to your gate, I should have been in danger (of continuing in my ignorance), and been laughed at for long in the schools of our great System."

北海海神若说:"对井底的青蛙不能谈论大海,这是受空间的限制;不可以对夏天的小虫子谈论寒冰,这是受时间的限制;不可以对浅陋偏执的人谈论道,这是受教育程度的限制。现在你从河岸来,看到大海,才知道自己的微小,可以和你一起谈大道理了。天下的水流,没有什么比得上大海,上万的河流之水流入大海,不知什么时候停止,大海却总也不满;水从海的出口倾泻而出,不知什么时候停止,大海的水却不见少;春季、秋季海水都不见变化,旱涝对它也没有影响。这是因为它的蓄水之多超过其他的江河,不可以用数量来计算。然而我并没有因此而自大,因为我从天地得到形体,从阴阳得到元气,我在天地之间,好像是在大山里的小石块、小树,我知道自己的渺小,又怎么能以此自夸呢!想想天下在天地之间,不像是大泽里的一个蚁窝吗?想想中原地区在广大的天下,不像是大粮仓里的一粒米吗?物体名称的数量有万种,人只是其中一种;人们聚集在九州里,谷物在这里生长,船车在这里通行,人只是其中之一。人和万物相比,不像是马身上的毫毛末端吗?五帝所关心的,三王所争夺的,仁人所担忧的,贤才所操劳的,都在这儿了。伯夷推辞它得到了名声,仲尼谈论它显示自己的知识丰富,这大概就是他们的自满吧,不和你之前觉得自己的水多是一样吗?"

Ruo, (the Spirit-lord) of the Northern Sea, said, "a frog in a well cannot be talked with about the sea; —he is confined to the limits of his hole. An insect of the summer cannot be talked with about ice; —it knows nothing beyond its own season. A scholar of limited views cannot be talked with about the Dao; —he is bound by the teaching (which he has received). Now you have come forth from between your banks, and beheld the great sea. You have come to know your own ignorance and inferiority, and are in the way of being fitted to be talked with about great principles. Of all the waters under heaven there are none so great as the sea. A myriad streams flow into it without ceasing, and yet it is not filled; and afterwards it discharges them (also) without ceasing, and yet it is not emptied. In spring and in autumn it undergoes no change; It takes no notice of floods or of drought. Its superiority over such streams even as the Jiang and the He cannot be told by measures or numbers; and that I have never, notwithstanding this, made much of myself, is because I compare my own bodily form with (the greatness of) heaven and earth, and (remember that) I have received my breath from the Yin and Yang. Between heaven and earth I am but as a small stone or a small tree on a great hill. So long as I see myself to be thus small, how should I make much of myself? I estimate all within the four seas, compared with the space between heaven and earth, to be not so large as that occupied by a pile of stones in a large marsh! I estimate our Middle States, compared with the space between the four seas, to be smaller than a single little grain of rice in a great granary! When we would set forth the number of things (in existence), we speak of them as myriads; and man is only one of them. Men occupy all the nine provinces; but of all whose life is maintained by grain-food, wherever boats and carriages reach, men form only one portion. Thus compared with the myriads of things, they are not equal to a single fine hair on the body of a horse. Within this range are comprehended all (the territories) which the five Dis received in succession from one another; all which the royal founders of the three dynasties contended for; all which excited the anxiety of Benevolent men; and all which men in office have toiled for. Bo-yi was accounted famous for declining (to share in its government), and Zhongni was accounted great because of the lessons which he addressed to it. They acted as they did, making much of themselves;—therein like you who a little time ago did so of yourself because of your (volume of) water!"

河伯说："那么我重视广大的天下而轻视细小的毫毛末端，可以吗？"北海海神若说："不行。万物的量是没有穷尽的，时间是无休无止的，得与失没有不变的常态，事物的终结和起始也没有定因。有大智慧的人观察事物顾及远近，所以小不一定少，大的不一定就多，这是因为知道事物的量是不可穷尽的。验证明察古今历史，寿命久远却不感到厌倦，生命就在眼前却不会祈求长寿，这是因为知道时间是无穷无尽的；明察事物盈虚的规律，所以并不因得到了而喜欢，也不因失去而忧虑，这是因为得失的秉分是没有定准的；明白了生死过程就像一条平坦大道，所以活着不会过分喜欢，死了也不把它当作一种祸患，这是因为知道事物的终结和开始不是一成不变的。计算人们了解的，不如不了解的多；人们活着的时间，也远远少于不在人世的时间。用自己最小的存在，去探求极大的领域，自己就会迷乱而无所得。由此看来，又怎么知道毫毛的末端就可以判定是最细小的限度呢？又怎么知道广大的天地就足够称得上最大的地域呢？"

The earl of the He said, "Well then, may I consider heaven and earth as (the ideal of) what is great, and a point of a hair as that of what is small?" Ruo of the Northern Sea replied, "No, The (different) capacities of things are illimitable; time never stops, (but is always moving on); man's lot is ever changing; the end and the beginning of things never occur (twice) in the same way. Therefore men of great wisdom, looking at things far off or near at hand do not think them insignificant for being small, nor much of them for being great：—knowing how capacities differ illimitably. They appeal with intelligence to things of ancient and recent occurrence, without being troubled by the remoteness of the former, or standing on tiptoe to lay hold of the latter：—knowing that time never stops in its course. They examine with discrimination (cases of) fulness and of want, not overjoyed by success, nor disheartened by failure：—knowing the inconstancy of man's lot. They know the plain and quiet path (in which things proceed), therefore they are not overjoyed to live, nor count it a calamity to die：the end and the beginning of things never occurring (twice) in the same way. We must reckon that what men know is not so much as what they do not know, and that the time since they were born is not so long as that which elapsed before they were born. When they take that which is most small and try to fill with it the dimensions of what is most great, this leads to error and confusion, and they cannot attain their end. Looking at the subject in this way, how can you know that the point of a hair is sufficient to determine the minuteness of what is most small, or that heaven and earth are sufficient to complete the dimensions of what is most large?"

河伯说:"世上议论的人都说:'最细小的东西是无形的,最大的事物大得不可限定范围。'这真实可信吗?"北海海神若说:"从细小的事物看来,大的无穷无尽,在巨大的事物看来,小的都看不清楚。精细,是小中至小;庞大,是大中至大,所以大小不同又各有合宜之处,这是事物本来的态势。所谓的精细与粗大,只是限于有形的东西;无形的事物是不可从数量上来区分的。而不可限定范围的事物,是不能用数量来计算的。可以用语言谈论的,是事物粗浅的表象;可以用心意传达的,是事物精细的内在本质;不可言谈的,不可用心意领会的,也就不限于精微和粗浅的范围了。因此,品德高尚之人的行为,不会出于伤害他人,不会刻意赞赏他人给人以仁惠恩德;做事不为私利,不轻视守门人一类的卑贱职业;不争财物,不过分地推辞谦让;凡事不借助他人的力气,但也不提倡自食其力,也不轻视贪婪和污秽;行动不合流俗,又不主张过分地追求怪异;行为追随一般人,也不鄙视奸佞谄媚;世上的爵位高禄不值得他心动,杀戮羞耻也不会让他感受到侮辱;知道是非不能截然分开,小大没有明确的界限。听人说:'得道的人不求闻名通达,德行高尚的人不贪恋财物之得,清虚宁静的人忘怀一已。'这是约束自己而达到恰如其分的境界。"

The earl of the He said, "The disputers of the world all say, 'That which is most minute has no bodily form; and that which is most great cannot be encompassed;' —is this really the truth?" Ruo of the Northern Sea replied, "When from the standpoint of what is small we look at what is great, we do not take it all in; when from the standpoint of what is great we look at what is small, we do not see it clearly. Now the subtile essence is smallness in its extreme degree; and the vast mass is greatness in its largest form. Different as they are, each has its suitability; —according to their several conditions. But the subtile and the gross both presuppose that they have a bodily form. Where there is no bodily form, there is no longer a possibility of numerical division; where it is not possible to encompass a mass, there is no longer a possibility of numerical estimate. What can be discoursed about in words is the grossness of things; what can be reached in idea is the subtilty of things. What cannot be discoursed about in words, and what cannot be reached by nice discrimination of thought, has nothing to do either with subtilty or grossness. Therefore while the actions of the Great Man are not directed to injure men, he does not plume himself on his benevolence and kindness; while his movements are not made with a view to gain, he does not consider the menials of a family as mean; while he does not strive after property and wealth, he does not plume himself on declining them; while he does not borrow the help of others to accomplish his affairs, he does not plume himself on supporting himself by his own strength, nor does he despise those who in their greed do what is mean; while he differs in his conduct from the vulgar, he does not plume himself on being so different from them; while it is his desire to follow the multitude, he does not despise the glib-tongued flatterers. The rank and emoluments of the world furnish no stimulus to him, nor does he reckon its punishments and shame to be a disgrace. He knows that the right and the wrong can (often) not be distinguished, and that what is small and what is great can (often) not be defined. I have heard it said, 'The Man of Dao does not become distinguished; the greatest virtue is unsuccessful; the Great Man has no thought of self;' —to so great a degree may the lot be restricted."

阅读拓展
More Reading for Practice

《桃花源记》
原文、现代汉语译文和英文翻译参考

桃花源记
陶渊明

晋太元中，武陵人捕鱼为业。缘溪行，忘路之远近。忽逢桃花林，夹

岸数百步,中无杂树,芳草鲜美,落英缤纷。渔人甚异之。复前行,欲穷其林。

　　林尽水源,便得一山。山有小口,仿佛若有光。便舍船,从口入,初极狭,才通人。复行数十步,豁然开朗。土地平旷,屋舍俨然。有良田美池桑竹之属;阡陌交通,鸡犬相闻。其中往来种作,男女衣着,悉如外人;黄发垂髫,并怡然自乐。

　　见渔人,乃大惊,问所从来,具答之。便要还家,设酒、杀鸡,作食。村中闻有此人,咸来问讯。自云先世避秦时乱,率妻子邑人来此绝境,不复出焉,遂与外人间隔。问今是何世？乃不知有汉,无论魏晋。此人一一为具言所闻,皆叹惋。余人各复延至其家,皆出酒食,停数日,辞去。此中人语云:"不足为外人道也。"

　　既出,得其船,便扶向路,处处志之。及郡下,诣太守,说如此。太守即遣人随其往,寻向所志,遂迷不复得路。

　　南阳刘子骥,高尚士也,闻之,欣然规往,未果,寻病终。后遂无问津者。

现代汉语译文参考

　　东晋太元年间,武陵郡有个人以打鱼为生。他顺着溪水行船,忘记了路程的远近。忽然遇到一片桃花林,生长在溪水的两岸,长达几百步,中间没有别的树,花草鲜嫩美丽,落花纷纷散在地上。渔人对此(眼前的景色)感到十分诧异,继续往前行船,想走到林子的尽头。桃林的尽头就是溪水的发源地,于是便出现一座山,山上有个小洞口,洞里仿佛有点光亮。于是他下了船,从洞口进去了。起初洞口很狭窄,仅容一人通过。又走了几十步,突然变得开阔明亮了。(呈现在他眼前的是)一片平坦宽广的土地,一排排整齐的房舍。还有肥沃的田地、美丽的池沼,桑树竹林之类的。田间小路交错相通,鸡鸣狗叫到处可以听到。人们在田野里来来往往耕种劳作,男女的穿戴,跟桃花源以外的世人完全一样。老人和小孩们个个都安适愉快,自得其乐。村里的人看到渔人,感到非常惊讶,问他是从哪儿来的。渔人详细地做了回答。村里有人就邀请他到自己家里去(做客),设酒杀鸡做饭来款待他。村里的人听说来了这么一个人,就都来打听消息。他们说他们的祖先为了躲避秦时的战乱,领着妻子儿女和乡邻来到这个与人世隔绝的地方,不再出去,因而跟外面的人断绝了来往。他们问渔人现在是什么朝代,他们竟然不知道有过汉朝,更不必说魏晋两朝了。渔人把自己知道的事一一详尽地告诉了他们,听完以后,他们都感叹惋惜。其余的人各自又把渔人请到自己家中,拿出酒饭来款待他。渔人停留了几天,向村里人告辞离开。村里的人对他说:"我们这个地方不值得对外面的人说啊。"渔人出来以后,找到了他的船,就顺着原路回去,处处都做了标记。到了郡城,到太守那里去说,报告了这番经历。太守立即派人跟着他去,寻找以前所做的标记,终于迷失了方向,再也找不到通往桃花源的路了。南阳人刘子骥,是个志向高洁的隐士,听到这件事后,高兴地计划前往。但没有实现,不久因病去世了。此后就再也没有问桃花源路的人了。

英文译文(translated by 林语堂)

The Peach Colony
Tao Yuanming

During the reign of Taiyuan of Chin, there was a fisherman of Wuling. One day he was walking along a bank. After having gone a certain distance, he suddenly came upon a peach grove which extended along the bank for about a hundred yards. He noticed with surprise that the grove had a magic effect, so singularly free from the usual mingling of brushwood, while the beautifully grassy ground was covered with its rose petals. He went further to explore, and when he came to the end of the grove, he saw a spring which came from a cave in the hill. Having noticed that there seemed to be a weak light in the cave, he tied up his boat and decided to go in and explore. At first the opening was very narrow, barely wide enough for one person to go in. After a dozen steps, it opened into a flood of light. He saw before his eyes a wide, level valley, with houses and fields and farms. There were bamboos and mulberries; farmers were working and dogs and chickens were running about. The dresses of the men and women were like those of the outside world, and the old men and children appeared very happy and contented. They were greatly astonished to see the fisherman and asked him where he had come from. The fisherman told them and was invited to their homes, where wine was served and chicken was killed for dinner to entertain him. The villagers hearing of his coming all came to see him and to talk. They said that their ancestors had come here as refugees to escape from the tyranny of Tsin Shih-huang (builder of Great Wall) some six hundred years ago, and they had never left it. They were thus completely cut off from the world, and asked what was the ruling dynasty now. They had not even heard of the Han Dynasty (two centuries before to two centuries after Christ), not to speak of the Wei (third century A. D.) and the Chin (third and fourth centuries). The fisherman told them, which they heard with great amazement. Many of the other villagers then began to invite him to their homes by turn and feed him dinner and wine. After a few days, he took leave of them and left. The villagers begged him not to tell the people outside about their colony. The man found his boat and came back, marking with signs the route he had followed. He went to the magistrate's office and told the magistrate about it. The latter sent someone to go with him and find the place. They looked for the signs but got lost and could never find it again. Liu Tsechi of Nanyang was a great idealist. He heard of this story, and planned to go and find it, but was taken ill and died before he could fulfill his wish. Since then, no one has gone in search of this place.

练习
Exercises

练习1 翻译课文

练习2 将下列段落先翻译成现代汉语再翻译成英语

养生主

吾生也有涯，而知也无涯。以有涯随无涯，殆已；已而为知者，殆而已矣。为善无近名，为恶无近刑。缘督以为经，可以保身，可以全生，可以养亲，可以尽年。

庖丁为文惠君解牛，手之所触，肩之所倚，足之所履，膝之所踦，砉然向然，奏刀騞然，莫不中音。合于《桑林》之舞，乃中《经首》之会。文惠君曰："嘻，善哉！技盖至此乎？"庖丁释刀对曰："臣之所好者道也，进乎技矣。始臣之解牛之时，所见无非全牛者。三年之后，未尝见全牛也。方今之时，臣以神遇而不以目视，官知止而神欲行。依乎天理，批大郤，导大窾，因其固然，技经肯綮之未尝，而况大軱乎！良庖岁更刀，割也；族庖月更刀，折也。今臣之刀十九年矣，所解数千牛矣，而刀刃若新发于硎。彼节者有间，而刀刃者无厚；以无厚入有间，恢恢乎其于游刃必有余地矣，是以十九年而刀刃若新发于硎。虽然，每至于族，吾见其难为，怵然为戒，视为止，行为迟。动刀甚微，謋然已解，如土委地。提刀而立，为之四顾，为之踌躇满志，善刀而藏之。"文惠君曰："善哉！吾闻庖丁之言，得养生焉。"

注释
Notes

（1）庖（páo）丁：名丁的厨工。先秦古书往往以职业放在人名前。文惠君：即梁惠王，也称魏惠王。解牛：宰牛，这里指把整个牛体开剥分剖。

（2）踦（yǐ）：支撑，接触。这里指用一条腿的膝盖顶牛。

（3）砉（huā）然：砉，又读 xū，象声词。砉然，皮骨相离的声音。向，通"响"。

（4）騞（huō）然：象声词，形容比砉然更大的进刀解牛声。

（5）桑林：传说中商汤时的乐曲名。

（6）经首：传说中尧乐曲《咸池》中的一章。会：指节奏。以上两句互文，即"乃合于桑林、经首之舞之会"之意。

（7）嘻：赞叹声。

（8）盖：通"盍（hé）"，何，怎样。

（9）进：超过。

（10）官知：这里指视觉。神欲：指精神活动。

（11）天理：指牛生理上的天然结构。

（12）批大郤：击入大的缝隙。批：击。郤：空隙。

（13）导大窾（kuǎn）：顺着（骨节间的）空处进刀。

（14）因：依。固然：指牛体本来的结构。

（15）技经：犹言经络。技，据清俞樾考证，当是"枝"字之误，指支脉。经，经脉。肯：紧附在骨上的肉。綮（qìng）：筋肉聚结处。技经肯綮之未尝，即"未尝技经肯綮"的宾语前置。

（16）軱（gū）：股部的大骨。

（17）割：这里指生割硬砍。

（18）族：众，指一般的。

（19）折：用刀折骨。

（20）发：出。硎（xíng）：磨刀石。

（21）节：骨节。间：间隙。

（22）恢恢乎：宽绰的样子。

（23）族：指筋骨交错聚结处。

（24）怵（chù）然：警惧的样子。

（25）謋（huò）：象声词。骨肉离开的声音。

（26）委地：散落在地上。

（27）善：通"缮"，修治。这里是擦拭的意思。

（28）养生：指养生之道。

现代汉语译文参考

我们的生命是有限的，而知识却是无限的。以有限的生命去探求无限的知识，会令人体乏神伤。既然如此还要不停地追求知识，可真是危险了！做了善事却不贪图名声，做了恶事却不至于受刑罚。遵从自然的中正之道并把它作为常法，这就可以护卫自身，就可以保全天性，就可以养父母，就可以终享天年。

有个名叫丁的厨师给文惠君杀牛，分解牛体时手接触的地方、肩倚靠着的地方、脚踩的地方、膝盖抵住的地方所发出的声响，以及进刀时发出的声音，无不合乎美妙的音乐韵律，符合《桑林》舞曲的节奏，又合于《经首》乐曲的乐律。文惠君说："嘻，妙呀！技术怎么会达到如此高超的地步呢？"厨师放下刀，回答说："我所喜好的是道，比起一般的技术、技巧更进一层。我开始杀牛时，所看见的没有不是一头整牛的。几年之后，就不曾再看到整体的牛了。现在，我只用心神去接触而不必用眼睛去观幕，眼睛的感官作用停了下来而精神世界还在不停地运行。依照牛体自然的生理结构，劈开肌肉骨间大的缝隙，把刀深入那些骨节间大的空处，顺着牛体的天然结构去解制；从不曾碰到过经络结聚的部位和骨肉紧密连接的地方，何况那些大骨头呢！优秀的厨师一年更换一把刀，因为他们是在用刀割肉；普通的厨师一个月就更换一把刀，因为他们是在用刀砍骨头。如今我的这把刀已经使用十九年了，所宰杀的牛有上千头了，而刀刃仍像刚从磨刀石上磨过一样锋利。牛的骨节之间是有空隙的，而刀刃是几乎没有什么厚度的，用薄薄的刀刃插入有空的骨节间，对于牛刀的运转和回放来说那是多么宽而有余地呀。所以，使用了十九年，我的刀刃仍像刚从磨刀石上磨过一样。虽然这样，每当遇上筋骨聚结交错的地方，我看到难于下刀，为此而格外谨慎不敢大意，目光专注，动作迟缓，动刀十分轻微。牛体'霍霍'地全部分解开来，就像是一堆泥土堆放在地上。我于是提着刀站在那儿，为此而环顾四周，为此而踌躇满志，最后把刀擦拭干净收藏起来。"文忠君说："妙啊，我听了厨师这一番话，从中得到了养生的道理。"

译文参考
Translation for Reference

Nourishing the Lord of Life

There is a limit to our life, but to knowledge there is no limit. With what is limited to pursue after what is unlimited is a perilous thing; and when knowing this, we still seek the increase of our knowledge, the peril cannot be averted. There should not be the practice of what is good with any thought of the fame (which it will bring), nor of what is evil with any approximation to the punishment (which it will incur): an accordance with the Central Element (of our nature) is the regular way to preserve the body, to maintain the life, to nourish our parents, and to complete our term of years.

His cook was cutting up an ox for the ruler Wen Hui. Whenever he applied his hand, leaned forward with his shoulder, planted his foot, and employed the pressure of his knee, in the audible ripping off of the skin, and slicing operation of the knife, the sounds were all in regular cadence. Movements and sounds proceeded as in the dance of "the Mulberry Forest" and the blended notes of "the King Shou." The ruler said, "Ah! Admirable! That your art should have become so perfect!" (Having finished his operation), the cook laid down his knife, and replied to the remark, "What your servant loves is the method of the Dao, something in advance of any art. When I first began to cut up an ox, I saw nothing but the (entire) carcase. After three years I ceased to see it as a whole. Now I deal with it in a spirit-like manner, and do not look at it with my eyes. The use of my senses is discarded, and my spirit acts as it wills. Observing the natural lines, (my knife) slips through the great crevices and slides through the great cavities, taking advantage of the facilities thus presented. My art avoids the membranous ligatures, and much more the great bones. A good cook changes his knife every year; — (it may have been injured in cutting); — an ordinary cook changes his every month; — (it may have been broken). Now my knife has been in use for nineteen years; it has cut up several thousand oxen, and yet its edge is as sharp as if it had newly come from the whetstone. There are the interstices of the joints, and the edge of the knife has no (appreciable) thickness; when that which is so thin enters where the interstice is, how easily it moves along! The blade has more than room enough. Nevertheless, whenever I come to a complicated joint, and see that there will be some difficulty, I proceed anxiously and with caution, not allowing my eyes to wander from the place, and moving my hand slowly. Then by a very slight movement of the knife, the part is quickly separated, and drops like (a clod of) earth to the ground. Then standing up with the knife in my hand, I look all round, and in a leisurely manner, with an air of satisfaction, wipe it clean and put it in its sheath." The ruler Wen Hui said, "Excellent! I have heard the words of my cook, and learned from them the nourishment of (our) life."

相关成语及其引申义

（1）游刃有余——现代人使用它来比喻技术熟练高超，做事轻而易举。

（2）目无全牛——一般用来指技艺达到极其纯熟的程度，达到得心应手的境界。

（3）踌躇满志——悠然自得，心满意足的意思。踌躇，一般用于形容犹豫不决的样子；踌躇满志，则是指对自己取得的成就洋洋得意的样子。

（4）切中肯綮——切中，正好击中。肯綮，是指骨肉相连的地方，比喻最重要的关键。

切中肯綮是指解决问题的方法对，方向准，一下子击中了问题的要害，找到了解决问题的好办法。

（5）批郤导窾——批：击；郤：空隙；窾：骨节空处。从骨头接合处劈开，无骨处则就势分解。比喻善于从关键处入手，顺利解决问题。

（6）新硎初试——硎：磨刀石；新硎：新磨出的刀刃。像新磨的刀那样锋利。比喻刚参加工作就显露出出色的才干。亦作"发硎新试"。

（7）官止神行——指对某一事物有透彻的了解。

（8）善刀而藏——善：拭；善刀：把刀擦干净。将刀擦净，收藏起来。比喻适可而止，自敛其才。

（9）庖丁解牛——厨师解割了全牛。比喻掌握了解事物客观规律的人；技术纯熟神妙；做事得心应手。

练习3　将下列句子先翻译成现代汉语再翻译成英语

大学之道，在明明德，在亲民，在止于至善。

知止而后有定，定而后能静，静而后能安，安而后能虑，虑而后能得。

现代汉语译文参考

高等教育的目标在于阐明（明）我们本性中智慧的（明）道德力量（德），在于建设一个更新和更好的社会（字面意思指"人民"），在于使我们遵守最高的美德。

只有当一个人拥有了关于美德的标准，唯有此时，他才会拥有一个固定而明确的目的；而只有有了固定而明确的目的，唯有此时，他才能拥有宁静而安稳的心灵；只有有了宁静而安稳的心灵，唯有此时，他的灵魂才能心神静谧；只有有了心神静谧的灵魂，唯有此时，他才能致力于深刻而严肃的思考；而只有通过深刻而严肃的思考，唯有此时，一个人才能获得真正的修养。

译文参考
Translation for Reference

The object of a Higher Education is to bring out the intelligent moral power of our nature; to make a new and better society, and to enable us to abide in the highest excellence.

When a man has a standard of excellence before him, and only then, will he have a fixed and definite purpose; with a fixed and definite purpose, and only then, will he be able to have peace and tranquility of mind; with tranquility of mind, and only then, will he be able to have peace and tranquility of soul; with peace and serenity of soul, and only then, can he devote himself to deep, serious thinking and reflection that man can attain true culture.

参考答案

第一部分

第一单元

1. A	2. C	3. B	4. D	5. D	6. C	7. B
8. B	9. A	10. D	11. D	12. C	13. C	14. B
15. C	16. A	17. A	18. C	19. D	20. A	21. C
22. C	23. C	24. A	25. B	26. A	27. D	28. A
29. C	30. A	31. A	32. D	33. D	34. C	35. A

第二单元

1. D   2. A   3. C   4. C   5. B

第三单元

1. A   2. B   3. A   4. D   5. B

第四单元

1. C   2. D   3. A   4. B   5. C

第二部分

第一单元

Passage A

1. B   2. A   3. B   4. D   5. C

Passage B

1. B   2. C   3. D   4. C   5. A

第二单元

1. B   2. C   3. A   4. D   5. B

第三部分

第一单元

Passage A

1. C   2. D   3. A   4. C   5. B

Passage B

1. C   2. D   3. C   4. B   5. A

第三单元

Passage A

1. C   2. D   3. A   4. A   5. B

Passage B

1. B   2. D   3. D   4. B   5. D

第四单元

1. D   2. C   3. A   4. B   5. D

第四部分

第一单元

Passage A

1. B   2. D   3. C   4. D   5. D

Passage B

1. A   2. C   3. D   4. B   5. D

第二单元

Passage A

1. B   2. A   3. B   4. C   5. A

Passage B

1. A   2. C   3. B   4. B   5. C

参考文献

[1] Edwin Gentzler. 当代翻译理论 [M]. 上海：上海外语教育出版社，2004：194.

[2] Mona Baker. 换言之：翻译教程 [M]. 北京：外语教学与研究出版社，2000：17-18.

[3] Susan Bassnett. 翻译研究 [M]. 上海：上海外语教育出版社，2010：5.

[4] 蔡希勤. 庄子说 [M]. 北京：华语教学出版社，2006：143.

[5] 陈扬建. 逻辑学基础教程 [M]. 北京：科学出版社，2011：14.

[6] 戴文进. 科技英语翻译理论与技巧 [M]. 上海：上海外语教育出版社，2003：56.

[7] 方梦之，范武邱. 科技翻译教程 [M]. 上海：上海外语教育出版社，2015：31.

[8] 冯庆华. 文体翻译论 [M]. 上海：上海外语教育出版社，2002：178-182.

[9] 冯小梅，王华琴. 英语时文阅读进阶 [M]. 北京：中国石化出版社，2002：85-87.

[10] 高永照. 科技英语阅读文选 [M]. 北京：国防工业出版社，2007：247-249.

[11] 辜鸿铭译. 大学中庸（中英双语评述本）[M]. 北京：中华书局，2017：259.

[12] 何其莘，仲伟合，许钧. 外事笔译 [M]. 北京：外语教学与研究出版社，2009：99.

[13] 何其莘，仲伟合，许钧. 英汉视译 [M]. 北京：外语教学与研究出版社，2009：2.

[14] 何兆熊. 语用学 [M]. 上海：上海外语教育出版社，2011：8.

[15] 黄伯荣，廖序东. 现代汉语 [M]. 北京：高等教育出版社，1991：232.

[16] 黄洪雷. 形式逻辑 [M]. 北京：中国农业出版社，2012：13.
[17] 理雅各译. 庄子 [M]. 北京：中州古籍出版社，2016：108-114.
[18] 潘文国. 英汉语比较与翻译 [M]. 上海：上海外语教育出版社，2016：132.
[19] 乔萍. 散文佳作108篇 [M]. 南京：译林出版社，2002：364-367.
[20] 萧立明. 英汉比较研究与翻译 [M]. 上海：上海外语教育出版社，2010：113.
[21] 许钧，穆雷. 中国翻译研究（1949—2009）[M]. 上海：上海外语教育出版社，2009：329.
[22] 赵萱，郑仰成. 科技英语翻译 [M]. 北京：外语教学与研究出版社，2006：86.
[23] 中央编译局译. 中华人民共和国国民经济和社会发展第十三个五年规划纲要 [M]. 北京：中国编译出版社，2016：67-79.
[24] 周方珠. 翻译多元论 [M]. 北京：中国对外翻译出版公司，2004：178.
[25] 周方珠. 英汉翻译原理 [M]. 合肥：安徽大学出版社，1997：352.
[26] 朱跃. 语义论 [M]. 北京：北京大学出版社，2006：108.

图书在版编目(CIP)数据

农林英语阅读与翻译/李清编著.—合肥:合肥工业大学出版社,2019.9
ISBN 978-7-5650-4646-9

Ⅰ.①农… Ⅱ.①李… Ⅲ.①农业—英语—阅读教学—高等学校—教材②林业—英语—阅读教学—高等学校—教材③农业—英语—翻译—高等学校—教材④林业—英语—翻译—高等学校—教材 Ⅳ.①S

中国版本图书馆 CIP 数据核字(2019)第 198776 号

农林英语阅读与翻译

READING AND TRANSLATION FOR AGRICULTURE AND FORESTRY

	李 清 编著		责任编辑 陆向军
出 版	合肥工业大学出版社	版 次	2019 年 9 月第 1 版
地 址	合肥市屯溪路 193 号	印 次	2020 年 9 月第 1 次印刷
邮 编	230009	开 本	710 毫米×1010 毫米 1/16
电 话	综合编辑部:0551-62903028	印 张	15.75
	市场营销部:0551-62903198	字 数	260 千字
网 址	www.hfutpress.com.cn	印 刷	安徽昶颉包装印务有限责任公司
E-mail	hfutpress@163.com	发 行	全国新华书店

ISBN 978-7-5650-4646-9　　　　　　　　　　　定价:39.80 元

如果有影响阅读的印装质量问题,请与出版社市场营销部联系调换。